EMPower Plus
extending mathematical power

AF822217

Split It Up

More Fractions, Decimals, and Percents

TEACHER BOOK

TERC
Because math and science build futures

Donna Curry, Mary Jane Schmitt, Tricia Donovan, Myriam Steinback, and Martha Merson

TERC 2067 Massachusetts Avenue
Cambridge, Massachusetts 02140

Cover
Maskot/Getty Images

***EMPower* Authors**
Donna Curry, Mary Jane Schmitt, Tricia Donovan, Myriam Steinback, and Martha Merson

Contributors and Reviewers
Michelle Allman, Bev Cory, Veronica Kell, Pam Meader, Heidi Schuler, Chelsey Amber Shade, Sherry Soares, Myriam Steinback, and Susan Swinford

Technical Team
Production and Design Team: Valerie Martin, and Sherry Soares

Photos/Images
Valerie Martin, Martha Merson, Myriam Steinback, and Rini Templeton

EMPower™ was developed at TERC in Cambridge, Massachusetts. This material is based upon work supported by the National Science Foundation under award number ESI-9911410 and by the Education Research Collaborative at TERC. Any opinions, findings, and conclusions or recommendations expressed in this publication are those of the authors and do not necessarily reflect the views of the National Science Foundation.

TERC is a nonprofit made up of teams of math and science education and research experts dedicated to innovation and creative problem solving. At the frontier of theory and practice, TERC's work encompasses research, content and curriculum development, technology innovation, professional development, and program evaluation. TERC has a passion for social justice and strives to create level playing fields for all learners, reaching more than three million students every year.

https://www.terc.edu/empower

Printed in the United States of America
1 2 3 4 5 6 7 8 9 10

ISBN 978-1-7373998-0-3

Send all *EMPower* content inquiries to:

Adult Numeracy Center at TERC
empower@terc.edu

Adult Numeracy Center

Dedication

We dedicate this revision to Mary Jane Schmitt and Tricia Donovan, outstanding educators, with gratitude for their vision and commitment.

Acknowledgments

Many friends and family members supported teachers' and *EMPower*'s efforts. Thank you all who made timely and substantive contributions to the first and second editions. We appreciate the encouragement and advice from our many colleagues. In addition, we thank the National Science Foundation, MetLife Foundation, and Fund for Public Schools, as well as TERC and the Education Research Collaborative at TERC for its support.

We are indebted to every adult student who participated in the piloting of *EMPower*. *EMPower* authors thank the teachers who were part of original pilot testing in AZ, ME, MA, NJ, NY, PA, RI, and TN in 2001–2004, as well as NYC teachers who participated in the pilot of the revisions in 2012–2013. The honest feedback and suggestions for what worked and what did not work were invaluable.

Contents

Getting Started with *EMPower Plus*

Background on *EMPower*

Extending Mathematical Power (EMPower) was the first math series that integrated mathematics education reform for educators, adult learners, and out-of-school youth. *EMPower* was designed especially for those students who return for a second chance at education by enrolling in remedial and adult basic education programs, high school equivalency programs, and developmental programs at community colleges. However, the series is appropriate for a variety of other settings as well, such as high schools, workplaces, and parent and paraprofessional education programs. *EMPower* builds interest and competency in mathematical problem solving and communication. The series serves as a model for a cohesive mathematics curriculum that offers content consistent with research and standards, including but not limited to the College and Career Readiness Standards and the Common Core Standards for Mathematical Practice. The series fosters a pedagogy of learning for understanding; it embeds teacher support and is transformative yet realistic for multi-level classrooms.

What's New in the 2nd Edition

This edition includes three fully updated books for students and teachers. Although *EMPower* users will recognize many of the activities from the first edition, we have also added new lessons and several opportunities for students to examine notation and algorithms (or methods) for solving problems with four basic operations. In response to the increased mathematical rigor of the College and Career Readiness Standards for Adult Education and the new high school equivalency tests, the three updated *EMPower Plus* books—*Everyday Number Sense, Using Benchmarks*, and *Split It Up*—help students build a foundation of number and operation sense for algebraic thinking. The work of cognitive psychologists and mathematics education researchers is starting to show what adult math educators have long suspected: attention to conceptual understanding with opportunities to notice and talk about patterns and strategies increases students' flexibility with number and problem solving. In these updated books, *EMPower* users will notice more opportunities for students to predict the results of operations and to see the connections between operations and their inverses (multiplying undoes division; square roots undo the effect of squaring). The excitement of learning math is in finding the meaning and the sense behind moves that once seemed magical or random. Recognizing mathematical properties at work has multiple benefits. Properties like the commutative property of addition and multiplication or the multiplicative identity make it possible for students to solve problems with understanding and with fewer errors. Recognizing the properties can lead students to value the strategies they use on a daily basis. Through *EMPower Plus* lessons, educators who have never thought much about operations, mental math, visual models, or benchmark numbers have opportunities to identify and encourage sturdy and reliable methods.

The Point of *EMPower*

EMPower consistently challenges students and teachers to extend their ideas of what it means to do math. The goal of *EMPower* is to help learners manage the mathematical demands of life by connecting situations and problems to mathematical principles. Situations include not only managing finance and commerce, but also interpreting news stories, applying health information, and facilitating family learning. *EMPower* is meant to be foundational, targeted to adults and young adults who test in the 4th-7th grade range who often have both areas of strength and gaps in their understanding. *EMPower* lessons introduce important mathematical ideas. They invite learners to identify patterns and to make connections. The lessons offer opportunities for students to explain their thinking and to justify their reasoning. Rather than focus on extensive, rote practice, *EMPower*'s main focus is to build learners' conceptual understanding, a critical platform from which they can explore more advanced concepts needed for future educational and career success.

In the following sections, you will read how *EMPower* shifts the culture of the classroom and how the lessons embody the College and Career Readiness Standards as well as the Common Core Standards for Mathematical Practice. The introduction provides an overview of the *EMPower* series as well as Frequently Asked Questions and Answers on both facilitation and the math content focus (number and operation sense with fractions) of the lessons within this book.

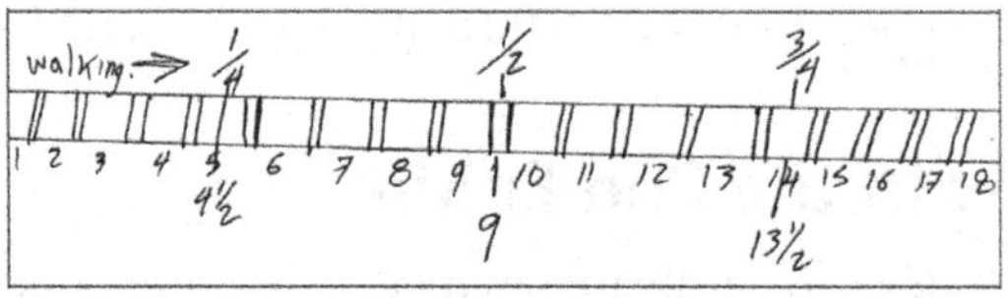

A student marks landmarks on an 18-block walk.

A Focus on Pedagogy and the Culture of the Classroom

Mathematics is meaningful within a social context. While mathematical truths are universal, the meaning and relevance of numbers change according to the setting and culture. *EMPower* classrooms become places where math ideas, strategies, hunches, and solutions are shared and discussed. The *EMPower* materials ask students to:

- Work collaboratively with others on open-ended investigations;
- Share strategies orally and in writing;
- Justify answers in multiple ways;
- Enter into and solve problems in various ways.

Key features of *EMPower* activities provide teachers with:

- Clear mathematical goals related to essential mathematics;
- Contexts that are engaging, challenging, and useful for adolescents and adults;
- Opportunities to strengthen learners' mathematical language and communication skills through productive struggle; and
- Puzzling dilemmas and problems that spark students' interest and motivate them to seek solutions.

These features make *EMPower* a resource for preparing students for tests of high school equivalency and community college coursework.

Students and teachers who experienced a traditional math education may find the expectations of *EMPower* take some getting used to. The chart highlights the contrast.

A rule-based approach emphasizes	*EMPower's* approach emphasizes
Computation, usually calculations by hand, with paper and pencil.	Mental math, visual models, and estimation using benchmark numbers, supported by a calculator when the numbers get unwieldy.
Procedures, often with limited attention to understanding; students practice following a given set of steps and then applying them to problems.	Conceptual understanding, sense-making, building on what students know, identifying patterns, and solving problems based on real or realistic contexts, all strengthened by connecting words, symbols, and visual models.
Completing the computational procedure correctly as evidence of understanding.	Being able to view the problem in a variety of ways, being able to communicate what the problem means or provide an example, and knowing what to expect as a sensible answer.
Applying the "correct" computational procedure to a problem.	Explaining and justifying their thinking, using strategies that illustrate flexibility and creativity when solving problems.

The Mathematical Practices and *EMPower*

EMPower's math background sections and unit introductions reference research in both K-12 and adult learning. In implementing *EMPower* lessons, teachers will foster the eight practices described in the Common Core and College and Career Readiness Standards for Mathematical Practice:

1. Make sense of problems and persevere in solving them.
2. Reason abstractly and quantitatively.
3. Construct viable arguments and critique the reasoning of others.
4. Model with mathematics.
5. Use appropriate tools strategically.
6. Attend to precision.
7. Look for and make use of structure.
8. Look for and express regularity in repeated reasoning.

The Practices describe independent, proficient mathematical thinkers. The essence of the Practices is visible and audible in *EMPower* classes as students:

- generalize, to explain their reasoning;
- use tools such as number lines, arrays, fraction strips, Pattern Blocks, and calculators;
- work with estimates, rounding, and place value;
- use the structure of numbers, e.g., breaking numbers into 10's and 1's, to solve problems;
- examine algorithms, generalizing from patterns to form rules, reasoning, and justifying;
- make observations about notation and operations that help them solve problems more efficiently.

Overview of EMPower Units

Features of the Teacher Book

FACILITATING LESSON **1**

Countries in Our Closets

Where were your clothes made?

Synopsis

This is the first of a set of lessons on categorical data in which students collect, examine, sort, and organize their own data, and then compare their data to another data set.

1. Students bring to the class data on the different geographic locations where their clothes were made.
2. The class combines its data into one frequency graph, and students make several statements about it.
3. Working in pairs, students compare their class graph to data from another class.
4. Pairs of students each create a new frequency graph, organizing the class data by continent. Students notice which continent is most prevalent.
5. Discussion centers on comparing country and continent graphs, closing with the question How does the story change when the graph's categories change?

Objectives

- Use a frequency graph to organize data
- Identify the story that the data tell
- Compare data from various samples
- Change the categories and articulate the change in the story

EMPower™ Facilitating Lesson 1: Countries in Our Closets 15

Each lesson opens with a **synopsis** of the lesson activities.

Objectives summarize the skills students should gain from the lesson.

Materials/Prep

- Grid paper
- Paper strips for posting statements
- Post-it® Notes
- Rulers

As an aid for determining in which continent the countries are located, have on hand a world atlas such as *The New Children's Illustrated Atlas of the World,* by Keith Lye (Philadelphia: Courage Press, 1999). For a listing of countries in Asia, see *Resource: Countries in Asia, Blackline Master 4.*

Make sure you have assigned *Countries in Our Closets* (*Student Book,* p. 7). Collect and bring in about 20 items of clothing in case your students have not collected their own data. (See *Facilitation,* p. 25, for how to handle this.)

Prepare a transparency or make a copy of the graph "Countries in Our Closets, Springfield, MA" (*Student Book,* p. 10).

Opening Discussion

Ask students:

Are clothes made in this city? In this country? Why or why not?

Why would anyone care where his or her clothes were made?

Acknowledge that we should know something about where our clothes come from, given how clothing is worn next to our skin. There are also tremendous economic and political issues at stake as manufacturing jobs are outsourced outside of the United States.

Heads Up!

If students do not bring in data, see *Facilitation,* p. 25.

You have looked in your closets, checked the tags on some of your clothes, and collected information. What did you learn?

Which country do most of the clothes you checked come from?

Post the country names in a place where you can easily refer to them later.

Introduce the terms **data set** and **mode.** Each student's list is a data set, and each data set could have a mode—the country name that occurs most frequently.

Vocabulary note: If students are searching for words to express their thoughts, offer terms and concise definitions. Students can keep track of these in their student books. For example, **data** are the information collected. The data that are collected make up the **sample.** The whole group from which information could be collected is the **population.** See *Lesson 1 Commentary,* p. 24, for more information.

16 Many Points Make a Point: Data and Graphs EMPower™

The **Materials/Prep** feature tells teachers what materials they will need to prepare for the day's activities.

The **Opening Discussion** gives teachers ideas for engaging students at the beginning of the lesson.

Heads Up! alerts teachers to helpful hints and things to look for in facilitating the lesson.

Speech bubbles and bold text suggest statements and questions teachers can use to spark and guide class discussion.

Summary Discussion

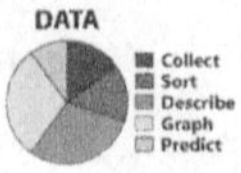

Finish the lesson by posing the following questions. (Students may not have ready answers at this point, but the questions will be a thread through the rest of the lessons.)

What stories can be told from this set of data (or what questions can be asked and answered)?

How does organizing the information into different categories impact how the data may be interpreted?

Who would be curious about what the people in this class eat?

Direct student to the *Reflections* questions (*Student Book*, p. 145).

Practice

Take a Sample, p. 24
For practice choosing a representative sample.

Thirsty and *Ouch!*, pp. 26 and 28
For practice organizing information and making statements.

Extension

Friends and Family Drink Too, p. 30

Data Collection, p. 31
Students begin recording what they eat for a project they will do in a later lesson. For now, they just write down all the foods they eat throughout the day.

Test Practice

Test Practice, p. 32

Looking Closely

Observe whether students are able to

Organize data into consistent categories

Do students group similar foods together? Make sure students are moving index cards around and that they are writing down the category names. If students seem stuck, have them ask you specific questions.

If you see foods categorized incorrectly, ask a question such as "Do all group members agree with the categories?" Encourage students to think, question, and defend their reasoning. If students use overlapping categories such as "cereal" and "grain," present an example close to their own experience. Say, for instance, "It is

EMPower™ Facilitating Lesson 2: Most of Us Eat... 33

A **Summary Discussion** guides the conclusion of the lesson.

Practice, Extension, and **Test Practice** sections list and describe choices for additional work.

Math Inspections are an opportunity to examine a mathematical convention or structure, to name a pattern, or to test an idea.

Looking Closely draws the teacher's attention to the goals and suggests teaching strategies.

LESSON 1 COMMENTARY

Rationale

The lesson takes students through the first steps of displaying data: collecting and organizing the information. As you work through this lesson, you and your students will notice that each time you categorize and recategorize the data set, you tell a different story.

Math Background

Data are information. This information may be numerical—e.g., salaries, test scores, heights, age, weight—as well as categorical—e.g., countries, foods we eat, types of product defects.

Most data are reported using statistics based on fractions and percents—one-third of pregnant mothers, half of native plants, etc. Commonly used, or benchmark, fractions and percents are often invoked to influence others to make decisions. When the headlines shout, "One-half of all pregnant women ...," readers tend to form opinions based on the data. What readers often neglect to consider is the size of the sample, where the data were collected, and how the data were organized.

The media generally report data that are based on a sample. It is rare for any group to have the time or resources to ask every person in the population to contribute information. The idea of sampling is to study a part in order to gain information about the whole. The sample—its size and characteristics—influences the data and the conclusions of the study.

A representative sample by definition includes representation across the population being surveyed. For example, a representative sample of the community would include individuals from across town, not just one neighborhood. Likewise, a representative sample across America would include individuals from many different states and regions, not just one or two. The size of the sample is important when inferences are made. For example, 30 samples could be sufficient for a population of several hundreds. The deciding factor for determining sample size is how confident you want to be about the inference.

It is not the concern of this unit to determine appropriate sample size, but it is important to note that statisticians use formulas derived from repeated surveys to decide sample size. Regardless, small samples are used in many studies. Factors such as availability of subjects or scarcity of time or financial resources might lead researchers to use a small sample and to base policy or claims upon the outcome.

Throughout this unit, it will be important to help students begin to think critically about data, whether the data are grouped into categories of information or into graphs. They will need to understand that a random sample will often yield different results from one that is not random. This topic resurfaces in *Lesson 2*.

24 Many Points Make a Point: Data and Graphs EMPower™

Math Background helps teachers deepen their understanding of mathematical concepts.

Context

Some students may know about *maquiladoras* in Mexican border towns, where women make clothes for very little money and with no benefits or environmental Occupational Safety and Health Administration (OSHA) workplace protections. CorpWatch (www.corpwatch.org) is one source for information on *maquiladoras*.

Facilitation

If students do not bring in data, or if their sample is too small, skip the second part of the *Opening Discussion*. Have available a pile of 20 clothing articles with labels. First, ask students to predict where the clothes were made. Post the list of their guesses. Note that it will be hard for them to answer this question unless they organize the information on the labels. Then divide up the 20 articles of clothing. Have students write the name of the country for each piece of clothing on a Post-it Note, one country name per note. Ask: "Where are most of our clothes made?" Then continue with the activity.

Making the Lesson Easier

Frequency graphs lend themselves to comparisons among categories. If students have little fluency stating comparisons, you may choose only to compare size, using terms like "greater," "fewest," or "less than." For students who are encountering data formally for the first time, the notion that collapsing data yields different stories may be difficult. Treat this lightly in the activity, and revisit such questions after students have more experience categorizing and recategorizing data in the homework and in *Lesson 2*.

Making the Lesson Harder

If your students can handle benchmark fractions and percents, get them to look critically at the data, including the source and sample size. You might ask:

If we asked another class what countries are in their closets, what do you think would happen to the categories? What if we asked the entire community?

How do you think your data would compare to data from another class of adult students in another community?

If students struggle with the idea of sample, you might try this: Have them each write their favorite color on a Post-it Note. If you have a small class, ask them to write the color on two Post-it Notes. Place all of the notes in a container. Have someone randomly (eyes closed) choose a few notes from the container and place them across a line to form a frequency graph. Ask the students how they think this sample compares to the actual total number of colors on notes in the container. You can have them do another frequency graph to compare the sample to the actual total.

Facilitating Lesson 1: Countries in Our Closets 25

The authors give ideas for **Making the Lesson Easier** and **Making the Lesson Harder.**

LESSON 1 IN ACTION

Alice articulates the mathematical principle behind compressed data.

I asked, "How did the change in categories affect what we noticed about the data?"

Alice answered, "Well, we keep losing information."

"How so?"

Patiently, Alice explained that when we started our work, every bit of data was visible. She added that we had lost details initially recorded. "At first, we knew every country in every person's closet and how many pieces of clothing came from that country. Then we combined the data, and we lost track of who had which countries. Then we did it by continent, and we lost track of all the countries."

Alice's realization quickly gained agreement from the rest of the class. After all, just the previous week a classmate had noted, "When you change the amount of data you look at, you find different things."

Sonia added her comment with increased conviction: "It is like politics. Politicians use a graph and tell you this is true, but you look at the graph, and it does not tell you everything."

Tricia Donovan
Pioneer Valley Adult Education Center, Northampton, MA

26 Many Points Make a Point: Data and Graphs EMPower™

In **Lesson in Action,** *EMPower* teachers share their classroom experiences.

Overview of EMPower Units

Features of the Student Book

LESSON **1**

Countries in Our Closets

Where were your clothes made?

In a world where information is at our fingertips, it is important to be able to make sense of it. What does it tell us? Why do we care?

Begin with your own clothes. Where were they made? Find out! You will collect **data** by reading the labels on your clothes. You will organize that data, make statements about it, and note where most of yo… were made.

You will then try organizing the data in a different way. H… the data in new groups change the story of the countries i… You will compare your class data to data of other classes. … be the same?

On Your Own

I think the country in which most of my clothes were mad…

Look at the labels of at least eight items of clothing. Make … track of the information you are collecting. Research ques… were your favorite clothes made?

EMPower™ Lesson 1: Countries in…

An **introduction** explains the purpose and content of the lesson.

Activity 1: Organizing the Data

Look at the data you collected about countries where your clothes were made.

Organize the data.

1. What strikes you about your data?

2. What is the **mode** for your data?

3. What is the mode for the class data?

8 Many Points Make a Point: Data and Graphs EMPower™

Vocabulary words are easy to identify in bold type.

The class **activity** comes next. Often students work with a partner or in a small group.

Practice pages invite students to use new skills.

Extension pages present a challenge or a new but related math idea.

A **Test Practice** that reflects the format of the test concludes each lesson.

Practice: Categorically Speaking

Sometimes we can organize categories into subgroups of other categories in order to better understand the data. For example, drug stores will often organize their aisles by category—skin care, beauty aids, and so on. Then the items in each of those aisles are also organized. This makes it easier to take inventory and place orders and for the customer to find products.

Can you think of other examples where information is categorized?

Use the chart below *or* create your own chart; draw a picture; make a list;

EMPow

Extension: Taking Inventory

Find a drawer or closet filled with many things. Use a frequency graph to show the contents. Start with six categories. Then do another frequency graph to show the same items, but this time use three categories.

For example:

My Medicine Cabinet

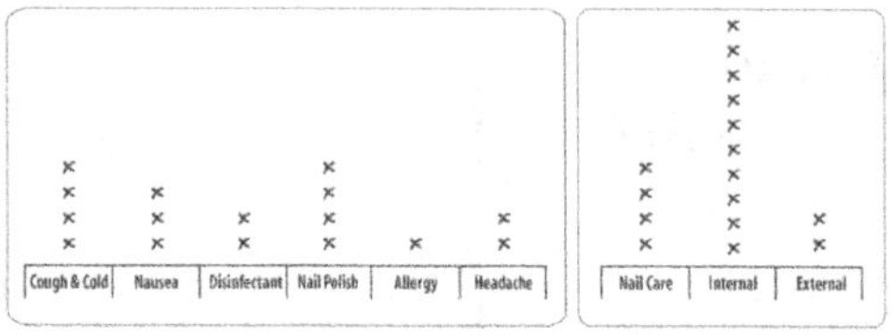

16 Many Points Make a Point: Data and G

Test Practice

1. Freda worked at a printing press. She kept a tally of all the defects she found in the books she was processing. Based on the frequency graph she created, shownbelow, which of the following statements is true?
   (1) There were more defects related to colors than any other defect.
   (2) There were twice as many torn pages as there were warped spines.
   (3) There were more warped spines than colors that bled.
   (4) Half of all the defects were related to colors that bled.
   (5) Half of all the defects were related to pages that were torn.

Book Defects Frequency Graph

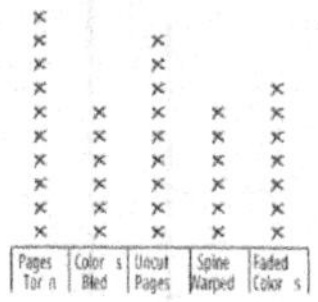

2. Tony, a tour guide, has been keeping a tally of visitors from different states. According to his tally, which of the following is *not* a true statement?
   (1) The fewest number of visitors came from Oregon.
   (2) There were as many visitors from Maine as there were from Georgia.
   (3) There were as many visitors from Georgia as there were from Oregon and Texas.
   (4) There were twice as many visitors from Florida as there were from Rhode Island.
   (5) There were twice as many visitors from Idaho as there were from Maine.

U.S. Tourist Frequency Graph

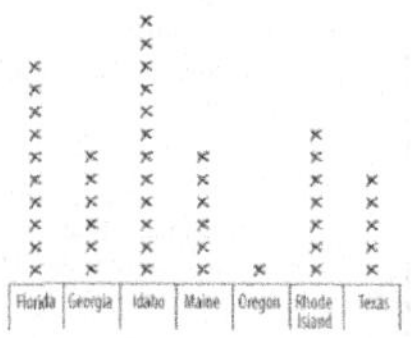

3. Clara is paid a commission on each of the large electronics devices that she sells. She tallied the different items she sold for the month. Based on her tally, what can she tell her boss about her sales?
   (1) She sold twice as many DVD players as she did TVs.
   (2) She sold twice as many DVD players as she did computers.
   (3) She sold more DVD players than she did TVs and computers combined.
   (4) She sold more TVs and VCRs than she did stereos and DVD players combined.
   (5) She sold half as many TVs as she did stereos.

Home Media Frequency Graph

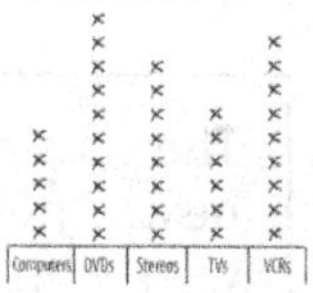

EMPower™ Lesson 1: Countries in Our Closets 17

Changing the Culture

Teachers who use *EMPower* often face the challenge of transforming the prevailing culture of their math classrooms. *EMPower* teachers have offered these ideas for facilitating this transition:

- **Set the stage.** Engage students in drafting and agreeing to ground rules. Explicitly state that this is a space for everyone to learn. As one teacher said, "We are in this together. Share, even if you do not think you are right. Whatever you add will be helpful. It lets us see how you are looking at things."
- **Group your students.** Match students whose learning styles and background knowledge complement each other. Ask questions such as, "How did it go to work together? How did everyone contribute?
- **Allow wait time.** Studies have shown that teachers often wait less than three seconds before asking another question. Students need time to think.
- **Sit down.** Watch students before interrupting to help them. Listen for logic and evidence of understanding. Follow the thread of students' thinking to uncover unconventional approaches. During discussions with the whole group, hand over the markers. Let students draw and make notes on the board.
- **Review written work.** Look beyond right and wrong answers to learn everything you can about what a student knows. Determine what seems solid and easy, as well as patterns in errors. If students are scattered, suggest ways they can organize their work. This is likely to lead to more efficient problem-solving and clearer communication.

Sequences and Connections

Any one of the *EMPower* books can stand alone, yet there are clear connections among them.

Some *EMPower* teachers alternate the books that focus on numbers with the books on geometry, data, and algebra. Particularly in mixed classes with some students at National Reporting System (NRS) Beginning Basic level or with older learners with minimal formal education, *Over, Around, and Within: Geometry and Measurement* works well as a starting point. The activities are concrete. Lessons introduce concepts like perimeter, area, and volume of two-dimensional shapes while keeping the computation focused on small whole numbers. Ratio and proportion unfold with scaffolding to insure that students can reason about rates and proportional relationships.

For teachers and students determined to become powerful problem-solvers with algebraic reasoning, begin with whole numbers and follow with fractions, decimals, and percents; ratio and proportion; and algebra. For this sequence, *Everyday Number Sense: Mental Math and Visual Models* is a starting point to develop whole number mental math skills and operation sense. *Using Benchmarks: Fractions and Operations* grounds students in part-whole relationships and operation sense with fractions. *Split It Up: More Fractions, Decimals, and Percents* continues to expand students' repertoire with part-whole ratios in fractions, decimals, and percents. *Seeking Patterns, Building Rules: Algebraic Thinking* builds upon the tools and relationships used in *Keeping Things in Proportion: Reasoning with Ratios*. The work with fractions and decimals is useful in describing approximate relationships between data sets in *Many Points Make a Point: Data and Graphs*.

Descriptions of the Books

Over, Around, and Within: Geometry and Measurement
Students explore the features and measures of basic shapes. Perimeter and area of two-dimensional shapes and volume of rectangular solids provide the focus.

EMPower Plus

Everyday Number Sense: Mental Math and Visual Models
Students solve problems and compute with whole numbers using mental math strategies with benchmarks of 1, 10, 100, and 1,000. Number lines, arrays, and diagrams support their conceptual understanding of number relationships and the four operations.

EMPower Plus

Using Benchmarks: Fractions and Operations
Students use the fractions $\frac{1}{2}$, $\frac{1}{4}$, $\frac{3}{4}$, and $\frac{1}{10}$; the decimals 0.1, 0.5, 0.25, and 0.75; and the percents 50%, 25%, 75%, and 100% as benchmarks to describe and compare all part-whole relationships. Students extend their understanding of the four operations with whole numbers to fractions. They decide upon reliable procedures for four operations with fractions.

EMPower Plus

Split It Up: More Fractions, Decimals, and Percents
Building on their command of common benchmark fractions, students build fluency finding tenths, 10%, and 1%. They work with decimals to the thousandths place. In analyzing the impact of the four operations on decimal numbers, they expand their repertoire and competence with part-whole relationships.

Seeking Patterns, Building Rules: Algebraic Thinking
Students use a variety of representational tools—diagrams, words, tables, graphs, and equations—to understand linear patterns and functions. They connect the rate of change with the slope of a line and compare linear with nonlinear relationships. They also gain facility with and comprehension of basic algebraic notation.

Keeping Things in Proportion: Reasoning with Ratios
Students use various tools—objects, diagrams, tables, graphs, and equations—to understand proportional and nonproportional relationships.

Many Points Make a Point: Data and Graphs
Students collect, organize, and represent data using frequency, bar, and circle graphs. They use line graphs to describe change over time. They use benchmark fractions and the measures of central tendency—mode, median, and mean—to describe data sets.

Frequently Asked Questions about Facilitation

Q: *Is there an ideal level for the* EMPower *series?*

A: The *EMPower Student Book* pages include situations and instructions that require some proficiency in written English. Students who test at NRS low and high intermediate levels or grades 4–7 grade level equivalency in mathematics are the best candidates for *EMPower.* Such students may have some familiarity with basic operations and know some number facts but might be unable to retain some procedures or perform them accurately and reliably—for instance, in the case of long division or of fraction operations. See the prerequisites for more information. Students who are higher level can benefit from *EMPower* if they have trouble getting started on a problem on their own, or if they are anxious and shut down when they see equations that look complicated. *EMPower* sets them up to be more independent, to test multiple solution paths, and to feel more confident in being flexible with numbers.

Q: *I have classes that are widely multi-level. Can this work?*

A: Many teachers see a wide range of levels within the group as an obstacle. Turn the range of levels to your advantage. Focus on students' representations (words, graphs, equations, sketches). This gives everyone the chance to see that answers emerge in several ways. Slowing down deepens understanding and avoids facile responses. Having calculators available can even the playing field. Implement the suggestions in *Making the Lesson Easier* and *Making the Lesson Harder* in the *Lesson Commentary* sections.

Q: *How do I deal with erratic attendance patterns?*

A: Uneven attendance can be disruptive. Students who miss class may feel disoriented; however, the lessons spiral back to the most important concepts. When the series circles back, students will have a chance to revisit concepts and get a toehold.

Q: *What do I do if I run out of time, and there is no way to finish a lesson?*

A: Each activity is important, but reviewing it is equally important. It is better to cut the activity short so there is time to talk with students about what they noticed. Maximize the time by selecting a student or group whose work you feel will add to the class's understanding to report their findings. Be conscious of when you are letting an activity go on too long because the energy is high. Fun is good, but be sure important learning is happening. If you like to give time in class to reviewing homework, and you want to hear from everyone in discussions, you will run out of time. Schedule a catch-up session every three or four lessons.

Q: *How do I respond to comments such as "Can't we go back to the old way?"*

A: Change is unsettling, especially for students who are accustomed to math classes where their job is to work silently on a worksheet solving problems by following a straightforward example. Be clear about the reasons why you have chosen to de-emphasize some of the traditional ways of teaching in favor of this approach. Ultimately, you may need to agree to some changes to accommodate students' input. Meanwhile, stick with the series. Reiterate for students what they have accomplished. When there is an "Aha!" moment, point it out.

Q: *My students don't have the time to go through a full curriculum. How can I convince students the value of using this program, even when their goal is to "get in and get out" as quickly as possible? Shouldn't I spend whatever time they have on a program designed to prepare them for college placement or high school equivalency tests?*

A: The National Center for Education and the Economy launched an intense study of the mathematics students need to be college and career ready. They determined that middle school math is vital for success in nine different programs offered at community colleges. They based their assessment on texts and exams from programs including nursing, accounting, and criminal justice. Though middle school math—fractions, decimals, percents, ratio, and proportion—are taught, they are not learned well. Teaching these concepts so that learners have a true foundation rather than a shaky, passing familiarity with a number of topics and procedures will enable students to meet their long-term goals.

Q: *My own math background is not strong. Will I be able to teach this series?*

A: Yes! Most teachers tend to teach the way they were taught. Adopting a different stance requires support, and the more types of support, the better. This series offers support in a few ways. The *Teacher Book* for each unit lists open-ended questions designed to keep the math on track. In the *Lesson Commentary* sections, the *Math Background* helps teachers deepen their understanding of a concept. In addition, the *Lesson in Action* sections provide examples of student work with comments that illuminate the underlying mathematics.

Consider joining the Adult Numeracy Network, http://adultnumeracynetwork.org, and attend your regional NCTM conference. You can find face-to-face and on-line course offerings through the Adult Numeracy Center at TERC, https://www.terc.edu/adultnumeracycenter/.

Split It Up Introduction

Overview

This book introduces fractions, decimals, and percents in a visual, conceptual way. It builds on operation and number sense in EMPower's *Everyday Number Sense* and *Using Benchmarks: Fractions and Operations*, which lay a foundation for work with percents and decimals. After extensive practice building fluency with the benchmark fractions halves, fourths, and eighths, this book offers learners an opportunity to examine procedures for solving problems with decimals.

This unit gives students many opportunities to make sense of rational numbers. Activities and practice pages encourage students to operate with benchmark fractions, decimals, and percents in a way that makes sense to them. Teachers tell us they love this unit because for the first time, their students are able to mentally calculate with percents by using the benchmarks 10% and 1%. By taking the time to establish the concepts of part-whole, a portion of an amount, and benchmark fractions, including tenths and hundredths, students develop a solid foundation for reasoning about data and continuing to make sense of numbers in their everyday lives.

Goals

As a result of completing the lessons in this book, students should be able to:

- Compare fraction, decimal, and percent amounts.
- Calculate to find the part, whole, or percent of an amount by using 10%, 1%, and their multiples.
- Decide on reliable methods for addition, subtraction, multiplication and division of decimals.
- Use mathematical symbols, real-world situations, and pictures to represent situations, estimate answers, solve problems, and justify solutions to decimal and percent problems.

Major Themes

The importance of developing number and operation sense

Many adults feel uncertain when they confront fractions, decimals, and percents. Perhaps the uncertainty stems from the fact that as children, they spent a lot of time learning the mechanics of arithmetic but did not develop a deep understanding of arithmetic principles and concepts. Stanislas Dehaene, in his book *The Number Sense*, believes the tradition of training children to be "little calculating machines that compute but do not think" results in a population of innumerate adults, who "are prompt in drawing hazardous conclusions based on reasoning that is mathematical only in appearance." Dehaene offers examples:

> "$\frac{1}{5} + \frac{2}{5} = \frac{3}{10}$ because $1 + 2 = 3$, and $5 + 5 = 10$."
>
> "$0.2 + 4 = 0.6$ because $2 + 4 = 6$."
>
> "0.25 is greater that 0.5 because 25 is greater than 5." (Dehaene, p. 137).

We share Dehaene's concern and see his examples as evidence of an individual's lack of a well-developed number sense with non-whole rational numbers.

But there is another concern we see, and that is adults' lack of operation sense with non-whole rational numbers. A common question in any math class is "What should I do—add, subtract, multiply, or divide?" We agree with the statement of the National Council of Teachers of Mathematics (NCTM) that to be proficient, learners need to "understand meanings of operations and how they relate to one another" (NCTM, 2000, p. 34). Some ideas people have about operations on whole numbers hold up when working with decimals or fractions. But others do not. In some cases, what people understand about whole numbers is helpful with fraction and decimal operations, and in other cases, previous knowledge should be questioned.

Operation sense includes understanding the relationships among the operations and the effect an operation will have on a pair of numbers (Huinker, p. 73). Lessons on operations are organized around dilemmas. The dilemmas are useful as opportunities to reason about rational numbers.

Another major idea is to learn operations not in isolation but with a sense of how they relate to one another: how addition relates to multiplication, how division relates to subtraction, and how multiplication relates to division, the commutativity of addition and multiplication only, and the order of operations.

Knowing how operations relate to one another gives a person a wider range of ways to approach any problem. Consider, for example, this problem: "How much do 2½ pounds of meat cost at \$3.00/lb?" Some people see it in terms of addition (\$3.00 + \$3.00 + \$1.50). Some see it in terms of multiplication (2.5 × \$3.00). The relationship between multiplication and repeated addition is why both approaches work.

"How much do two and one-half pounds of meat cost at \$3.00/lb?"

Some people see it in terms of addition: \$3.00 + \$3.00 + \$1.50

Some see it in terms of multiplication: 2.5 × \$3.00.

In addition, the lessons are designed to develop number sense for large and small numbers, bolstering students' understanding of place value. Numerate adults have strong number sense for common quantities. Many mathematics educators consider number sense to be the main objective of basic mathematics. Although there are varied definitions of number sense, most "focus on its intuitive nature, its gradual development, and ways in which it is manifested. The lessons build students' ability to "make sense of numerical situations" by developing the hallmark abilities of number sense: using numbers flexibly when mentally computing, estimating, judging number magnitude, and judging reasonableness of results (Markovitz and Sowder, p. 5).

In the unit that precedes this one (*Using Benchmarks: Fractions and Operations*), number sense is developed by providing students with opportunities to build a fluency with commonly used rational numbers (½'s, ⅓'s, ¼'s, and ⅛'s); and percents (the equivalents of those fractions). These everyday fractions, decimals, and percents serve as benchmarks for reasoning with less friendly rational numbers, and they continue to do so in this book.

Connecting symbolic notation, visual models, and real situations

The National Research Council's (NRC) report, *Adding It Up: Helping Children Learn Mathematics* (2001), summarizes the research on the development of children's mathematical proficiency. Their conclusion about rational numbers is that instructional programs that use "approaches that build on students' intuitive understanding and use objects or contexts that help students make sense of the operations offer more promise than rule-based approaches" (NRC, 2002, p. 416). We think this advice, though written for elementary-school mathematics educators, makes good sense for adult mathematics educators as well.

Therefore, this unit (like others in the series) is guided by NRC's recommendation:

> The curriculum should provide opportunities for students to develop a thorough understanding of rational numbers, their various representations including common fractions, decimal fractions, and percents, and operations on rational numbers. These opportunities should involve connecting symbolic representations and operations with physical or pictorial representations, as well as translating between various symbolic representations . (NRC, 2002, p. 416)

The perspective on how to build efficiency for operating with fractions, decimals, and percents is quite different in *EMPower* from that in most basic math workbooks for adults and adolescents. Traditional rule-based approaches focus on first spending a good amount of time teaching paper-and-pencil procedures (algorithms), followed by offering applications in the form of word problems to practice those procedures. *EMPower* emphasizes a different set of skills in a different sequence. Experience and research suggest this is more suited to the way confidently numerate adults make use of math in their everyday lives. In the *EMPower* books, students are encouraged to build upon what they know about operations and also to expand their repertoires. They justify reasoning with objects, diagrams, real-life situations, and their own number sense, which is grounded in the benchmark fractions, decimals, and percents instead of focusing on the memorization of rules.

Prerequisites

Split It Up: More Fractions, Decimals, and Percents builds on content in *Using Benchmarks: Fraction and Operations* in which students became familiar with the everyday benchmark fractions ½, ¼, ¾, and ⅛. *Split It Up* assumes that students have a solid grasp of those benchmarks and can confidently use them to find parts of amounts. For example, they can mentally calculate ¾ of $200. They can also use everyday fractions to approximate the fractional part of a total. For example, they are able to say that $70 is more than one-fourth of $200, but less than half of it. Ideally they can work from a part to find the whole: "If $15 is ¾ of the amount I owe, what is the total owed?"

In this unit, students efficiently and accurately use a variety of benchmarks to solve percent and decimal problems. They use 10% as a benchmark to mentally calculate multiples of 10%. They learn 1% and use combinations of 10% and 1% to figure out other percents of amounts. They examine decimal amounts in the hundredths and thousandths, deepening their understanding of place value. They apply what they know about the four basic operations to operations with decimals.

Use this unit if students have good working knowledge of the 2's and 10's multiplication facts, even if they struggle to recall other multiplication facts. In the service of thinking about fraction, decimal, and percent relationships, the investigations reinforce multiplication and division concepts and provide opportunities for students to practice multiplying and dividing.

Although it may appear basic, this unit builds students' understanding and ability to work with benchmark fractions, decimals, and percents in a way that makes sense to them. Some *EMPower* pilot teachers who were hesitant to use the unit repeatedly expressed surprise and enthusiasm for how much they saw students learn. They said student reliance on teacher direction decreased, and problems that were initially difficult for them (e.g., "What is the 15% tip on a $12.50 meal?") became simple mental calculations. By taking the time to establish the concepts of part-whole, a portion of an amount, and benchmark fractions, decimals, and percents, students will lay a foundation that will serve them well as they continue to make sense of numbers in their everyday lives.

The Flow of the Unit

This unit contains

- Facilitation notes for *Opening the Unit* and *Closing the Unit*
- 11 lessons

The unit offers many opportunities for students to build depth for the part-whole model and the portion of an amount model for fractions, decimals, and percents. Students use number lines as a tool to show their thinking.

During the first five lessons, the emphasis is on $\frac{1}{10}$, 10%, 1%, and their multiples and combinations. *Lessons 6* and *7* shift the focus to decimal hundredths and thousandths. In *Lessons 8*, *9*, and *10*, students operate with decimals. In *Lesson 11* they apply what they have learned throughout the unit

The lessons take on common errors with decimals (e.g., mistakes in lining up the decimal points for addition and subtraction; applying mis-remembered rules about moving the decimal places). Students are consistently asked to visualize the quantities and estimate the answers. Then students begin to probe the computational algorithms by reconstructing what they already know or remember and comparing those step-by-step procedures to arrive at a rule or generalization. In every case, the algorithms used to compute with decimals, fractions, and percents should make sense. Calculators are tools, not a replacement for thinking. Each lesson has one or more activities or investigations. Allowing time for opening and summary discussions (including time for student reflections) and assuming a thoughtful pace, most lessons will take two hours or more.

In *Lesson 1*, students

- Find one-tenth of a quantity.
- Identify multiple ways of representing one-tenth and relate them to visual models.

In *Lesson 2*, students

- Share ways to determine 10%, or $\frac{1}{10}$, of an amount.
- Determine a total, given 10% of it.

- Determine whether a variety of part-whole situations are more than, less than, or equal to 10%.

The lesson begins with the fraction $\frac{1}{10}$, which is the foundation for the unit and the way to build the connection between decimals and fractions. If students do not have ways to show one-tenth, review *Using Benchmarks: Fractions and Operations, Lesson 1: More Than, Less Than, or Equal to One-Half?* and *Lesson 2: Half of a Half?* (*Student Book*, pp. 7-43).

In *Lesson 3*, students

- Share strategies to determine multiples of 10% of a given amount.
- Base decisions involving percents on the fact that the whole of an amount or space equals 100%.
- Show percents with arrays of 50 and 100 squares.
- Name multiples of 10% with equivalent fractions.

In *Lesson 4*, students

- Find 1% and its multiples of three- and four-digit numbers.
- Compare 10% of one amount and 1% of another to articulate the effect of the size of the percent and the whole on the answer.

In *Lesson 5*, students

- Use multiples of 1% to find single-digit percents.
- Use multiples of 10% and 1% combined to find two-digit percents.

In *Lesson 6*, students

- Visualize decimal place value in the tenths and hundredths.
- Connect benchmark fractions to decimal equivalents to the hundredths.

In *Lesson 7*, students

- Explore the value of decimal places smaller than tenths and hundredths.
- Write decimals in expanded notation.
- Round decimals in the thousandths to the nearest 1, 0.1, 0.01.

In *Lesson 8*, students

- Build on their understanding of the meaning of addition and subtraction operations with whole numbers and fractions to generalize rules for adding and subtracting decimals.
- Further refine their sense of place value and its importance when working with decimals and percents.

In *Lessons 9* and *10*, students

- Conjecture about and decide upon reasonable procedures for multiplying and dividing with decimal numbers.
- Use visual models to support mathematical procedures for multiplying with decimal numbers.
- Demonstrate understanding of division with decimals in various ways.
- Apply the properties of arithmetic (e.g., commutative, distributive, associative) to multiplication with decimals.

In *Lesson 11*, students

- Practice using decimal numbers in real-life applications.

A Note on the Assessments and Pacing

Initial and Final Assessments

Split It Up opens and closes with assessment. In both cases, the *Teacher Book* provides multiple ways to gauge what students know. Teachers may spread the tasks out over multiple days or delay the written tasks of the pre-assessment particularly if students have just completed a school-administered placement test.

Ongoing Assessment

Much of the work students do will take place in small groups or pairs. This work must be evaluated to determine the extent to which students are comfortable with the main ideas and to diagnose difficulties, as well as to determine when to provide more challenging work. The *Looking Closely* section of each lesson focuses teachers' attention on the lesson's objectives and the corresponding observable behavior. Look for evidence of understanding and when a majority of students grasps concepts, move on. The material is spiraled so that students have multiple chances to take command of the material. As students gain skills, teachers will want to track their progress and communicate observations to them.

Frequently Asked Questions about the Math Content

Q: Why focus on meaning making? Don't students benefit most from repeated practice?

When meaning is lost, it is difficult to leverage the intuitive or to make common sense we can bring to problem solving, to recognize patterns, to generalize, and to make connections.

In *EMPower* books, students are encouraged to build upon what they know. They justify reasoning with objects, diagrams, real-life situations, and their own number sense, which is grounded in the benchmark fractions, decimals, and percents. In summarizing the research on children's mathematical proficiency, the National Research Council found that instructional programs for rational numbers that build on students' intuitive understanding and use of objects or contexts offer more promise than rule-based approaches" (NRC, 2001).

Q: Why the emphasis on developing number sense through mental math?

A: There are at least three reasons for the emphasis on developing mental math skills with whole numbers, everyday fractions, decimals, and percents: (1) mental math skills are a basis by which to judge the reasonableness of estimates; (2) mental math skills are a basis by which to judge the reasonableness of error-prone rule-based calculations and electronic entries; and (3) mental math strategies accentuate the structure of the number system—an important base for measurement as well as algebraic problem solving.

Numerate adults have options. They can be flexible in their approach to problems. For example, they might break apart numbers in different ways, seeing .36 as three-tenths and six-hundredths or as three times one-tenth and three times two-hundredths. Further, they might identify the quantity as slightly more than one-third or about halfway between .25 and .50. They can compare numbers to one another in an absolute way (How many more do I need to add to .25 to get 2?) as well as in a relative way (2 is how many times greater than .25?). Strong number sense means adults can use what they know to figure out things they don't know, such as mentally calculating half of one-fourth to figure out the value of 12.5%.

Q: What role does operation sense play?

A: Many adults confront math problems and find themselves uncertain which operation to use: addition, subtraction, multiplication, or division. Operation sense includes understanding the relationships among the operations, and the effect an operation will have on a pair of numbers (Huinker, 2002).

Operation sense also includes understanding the meanings and models of operations, the real-world situations they connect with, and the symbols that represent them. Adults' limited understandings of operations with whole numbers often leads to over-generalizations, such as:

- Multiplication makes things bigger.
- You cannot divide a small number by a larger number.
- To add decimals, line them up vertically, as with whole numbers right-justified.

EMPower's books on rational numbers present students with opportunities to reason about such generalizations.

Q: Why does having different models for operations matter?

A focus on the behavior of operations allows students to start in familiar territory of number and computation to progress to true engagement in the discipline of mathematics (Russell, Bastable, & Schifter, 2011). Recognizing problem types and testing or matching them to different models ultimately gives a person a wider range of ways to approach any problem.

Q: Why do students seem to fall apart in the face of fractions?

A: Proficiency with rational numbers requires a rich and integrated understanding of their various forms and meanings. Every rational number has multiple representations, and every rational number has several meanings. A common fraction has the potential for *infinite* labels: ¾ = 0.75 = 0.750 = 0.75000 and = 75%. In *EMPower*, students are asked early on to make sense of equivalencies. The part-whole model and the portion of an amount model for fractions are central to *Using Benchmark Fractions* and *Split It Up!* Other models surface in other books, e.g., in *Keeping Things in Proportion*, the concept of fractions representing comparative relationships is primary.

If students don't have a strong sense of forms and meanings as well as relative value of benchmark fractions like quarters and tenths, their ability to reason about fractions and decimals is limited (Givvin, Stigler, Thompson, 2011).

Q: Why focus on algorithms and properties of arithmetic?

A: Though an algorithm is any method that works, some methods undermine students' mathematical understanding because they short-cut meaning-making opportunities. Some algorithms conflict with a method learned early on. The emphasis here is examining sturdy methods that can be connected to a visual representation, so the meaning is kept intact.

In the 21st century, the reasons for reviewing arithmetic have less to do with teaching computation (shopkeeper math) and more to do with laying the foundation for understanding the structure of mathematics. Too often students manipulate elements in an algebraic equation without understanding why some moves are allowable and others are not. They become increasingly frustrated when equations and models become more complex. They have no foundation in the properties (e.g., commutativity, associativity, distributivity, identity, and inverse relationships) on which to build. The point is for learners to understand *why* what they are doing works because they will need to apply the skills they are learning broadly as we help them prepare for college and careers.

Split It Up Materials List

Lesson	Recommended materials	
Opening	Colored markers (at least 3 colors, to represent fractions, decimals, and percents) Easel paper Magazines and newspapers, such as *Time* and *USA Today*, 1 per student pair	Copies of the *Initial Assessment*, p. 133, 1 per student Copies of the *Initial Assessment Checklist*, p. 138, 1 per student (optional) Pennies for *Pass the Pennies*, 20 pennies per student (optional)
1	One box of 100 paper clips 100 pennies and 10 dimes (optional) Measuring tapes, in inches 3 X 5 index cards, at least 1 per student Ruler **The above materials are enough to set up one each of five different stations; for a large class, materials may be needed to set up duplicate stations.	Scissors (a few pairs would be helpful) Clear tape One quarter (25¢) 25 pennies for reference 4 stamps or a copy of *Blackline Master 1: Stamps*
2	Calculators	*Blackline Master 2: Match Cards*, 1 copy per student, plus 1-2 extra copies to cut up
3	Calculators Sticky notes Tape Copies of building blueprints (optional) Colored pencils or markers in at least 3 colors for student use *Blackline Master 11: 50-Block Grids*, 1-2 per student	Easel pad A large version of *Blackline Master 3: Two Out of Twenty* A large quantity of small objects (beads, tiles, toothpicks, paper clips, etc.), enough so that each student receives 3-10 of the same type of object
4	Calculators Colored markers Pennies or colored counters (if pennies are used, $10 of pennies in rolls is useful) Scissors Tape	Copies of *Blackline Master 4: One Percent on the Number Line*, 1 per student Copies of *Blackline Master 5: Money Strips*, at least 4 Copies of *Blackline Master 10: 100-Block Grids* (optional)
5	Calculators Colored pencils or markers Play money Sticky notes (optional) Copies of *Blackline Master 9: 100-Block Grid* 1 per pair or group	Actual sales slips showing item costs and sales tax (from more than one state if possible) Payroll stubs with names blacked out

Lesson	Recommended materials	
6	Calculators Meter stick A place value chart that can be posted	*Blackline Master 6: Ten Cards*, 1 per pair of students, plus an enlarged copy for the board A meter-long number line marked at 10 cm intervals
7	Calculators *Blackline Master 7: Thousandths Cards*, 1 cut-up set per student Envelopes, 1 per student, to keep Thousandths Cards in	10-sided die or *Blackline Master 8: 0-9 Cards*, 1 set or die per student or pair Copies of an electric bill or students' own electric bills (optional)
8	Calculators Fraction strips (optional)	
9	Items for counting such as chips or pennies, about 50 per student Fraction strips (from earlier lessons) Graph paper	Rulers *Blackline Master 10: 100-Block Grids*, 1-2 per small group Easel paper Markers
10	Calculators Rulers *Blackline Master 10: 100-Block Grids*	Fraction strips marked with equivalents
11	Full cereal box (if doing *Activity 2, Project 2*) Rulers (cm)	Stock market performance data (if doing *Activity 2, Project 4*)
Closing	Copies of *Final Assessment*, 1 per student Students' assignments representing their work in this unit	Copies of *Final Assessment Checklist*, 1 per student (optional)

Facilitating Opening the Unit: Split It Up!

What are some ways to split things up?

Synopsis

This lesson welcomes students to the unit by asking them to find examples of fractions, decimals, and percents in print matter—a magazine or a newspaper—and to select some they will use for basic mental computation. Then students complete the *Initial Assessment.*

1. Student pairs locate examples of fractions, decimals, and percents in newspapers and magazines. They select one of each type of number and mentally compute multiples and portions of those amounts.
2. Students take the *Initial Assessment* to determine what they already know or don't know about fractions, decimals, and percents.

Objectives

- Identify fractions, decimals, and percents in print materials
- Demonstrate ability to solve problems involving fractions, decimals, and percents

Materials/Prep

- Colored markers (at least three colors for fractions, decimals, and percents)
- Easel paper
- Magazines and newspapers (e.g., *Time, USA Today*), one per student pair
- Photocopies of the *Initial Assessment, Appendices*, p. 131, one per student
- Copies (one per student) of *Initial Assessment Checklist, Appendices*, p. 136 to record your observations for individuals (optional)
- Pennies for *Pass the Pennies*, 20 pennies per student (optional)

Opening Discussion

Begin by telling students that as they work with fractions, decimals, and percents, they can start looking for instances in which fractions, decimals, and percents appear in the news.

Tell students you are interested in knowing which form occurs most frequently and how they would solve problems with the numbers they find. Today is a day to demonstrate what they already know about working with fractions, decimals, and percents.

Activity 1: Newspaper and Magazine Search

Ask students to work with a partner, and refer them to *Activity 1: Newspaper and Magazine Search* (*Student Book*, p. 2). Distribute newspapers and magazines.

Review the directions:

> Search for examples of fractions, decimals, and percents. Circle them and mark them with an F for fraction, D for decimal, or P for percent.
>
> Pick one of each form that interests you: a fraction, a decimal, and a percent. For each one, mentally find twice as much, half as much, 10 times as much, and a tenth as much as the original amount. Fill in your chart.
>
> Emphasize that if students are not sure of the exact amount, they should make the best **estimate** they can.

As pairs search, listen to what they say. Is there any discussion or disagreement about categorizing the numbers (e.g., 3.2%)?

Headlines, news stories, and ads might distract students as they make sense of the numbers. Take some time to talk about what interests them, and encourage questioning of the reported numbers in a news story, but then move the activity along.

Bring the class together to address the frequency of each number form. Students share which they found more frequently—fractions, decimals, or percents—and discuss reasons why that might be.

Heads Up!

It might surprise students how few fractions show up in magazines and newspapers. If no fractions are located, supply one, such as 3/4, for students to use as their fraction form. See *Opening the Unit in Action* to read about how this activity played out in class.

End the activity by saying:

How is finding half as much of a number different from finding twice as much of it? How is it similar?

How is finding 1/10 of a number different from finding 10 times as much of it? How is it similar?

Ask for specific examples. Take note of how well students are able to find half. Are they able to do it accurately? If not, that is an indication that a review of *Using Benchmarks: Fractions and Operations* is necessary.

Activity 2: Initial Assessment

Distribute a copy of the *Initial Assessment* to every student. Begin with a preview of the questions. Remind students that this is an initial assessment so they are not expected to know all the answers. This is a way to learn what they know and don't yet know. Leave calculators aside for now.

Summary Discussion

Ask students to record their thoughts about the initial assessment, under *Reflections* (*Student Book*, p. 207) focusing on their strong points and points at which they were challenged.

Ask students to turn to p. 3 in their student books. Explain that the goal of this unit is to prepare students so they can solve any of these problems and others like them. They will use the methods they already know and explore new ways to look at problems.

Review the goals listed and ask students to use the space underneath the list to add their own goals. To prompt additions, you might ask:

What are you curious about?

Direct students to *Vocabulary (Student Book*, pp. 203-206) and encourage them to use their own words and examples to describe the terms. Ask if there are any words they would like to add to the list. Collect the *Initial Assessments*, and use the *Initial Assessment Checklist* to give students a sense of what you see as their strengths and which of their skills need development.

Looking Closely

Observe whether students are able to

Identify fractions, decimals, and percents in print materials

This task should be fairly straightforward. However, there may be some numbers, such as 9.5%, that raise questions about whether they are decimals, percents, or both.

What about numbers such as $4.37 billion and its value? This may be an opportunity to discuss how to write this number without the decimal point as $4,370,000,000.

As students search for fractions, the discussion of what a fraction is might arise. See *Opening the Unit in Action* (p. 7) for some examples of dilemmas involving fractions that came up in two classes.

Demonstrate ability to solve problems involving fractions, decimals, and percents

When calculating with percents, fractions, and decimals, look for a window into students' understanding of place value. Listen for half-remembered algorithms or methods they may have learned in other countries. Building on prior knowledge will increase the chances that students remember new material. Notice if students compare amounts to fraction equivalents (e.g., halves and quarters, or eighths (12.5%)).

Rationale

This is an assessment lesson, an opportunity to welcome students to the unit and for you to get an idea of what your students know about fractions, decimals, and percents through their comments and explanations. This lesson invites students to think about rational numbers and to connect the numbers they see in the world outside the classroom with their study of mathematics.

Math Background

The percents covered in this unit begin with 10%, 1%, and their multiples. The idea is that by finding 10% and 1%, we can use those numbers to find any other percent through strategies such as doubling, taking half, adding, or subtracting. For example, to find 17%, we can find 10%, double that and get 20%, and subtract 3% (1% x 3); or we can find 10%, then half again as much, or 5%, and then 2%. Students who can think of percents flexibly will successfully find any percent.

The fractions in this unit build on halves, quarters, and eighths explored in *Using Benchmarks: Fractions and Operations.* Students find one-tenth and use grid paper and arrays to reason about hundredths.

Decimals arise naturally as students make connections between what they know about percents and their corresponding decimal names. If students are familiar with using calculators and money, they have seen decimals, and the task now is to understand deeply what 0.01 means, for instance.

Students in pilot classes wondered why they couldn't just move the decimal over one place to find a percent. We found, however, that they were not clear about which direction to move the decimal—left or right. We further noticed that they did not know because this strategy was just "a rule" to them. The purpose of the lessons and activities in this unit is to help students understand the concept of finding a fraction, percent, or whole, and be able to select methods to do so that are appropriate to and useful in the circumstances they encounter.

Facilitation

Introduce the *Unit Goals* (*Student Book*, p. 3), if there is time. If not, return to them later. The goals might have more meaning at a later time when students have more familiarity with the concepts.

Making the Lesson Easier

The purpose of *Activity 1* is to see whether students can easily calculate ½ and 1⁄10 of an amount, as well as two and ten times an amount. That facility is expected in the unit. Here is a description of *Pass the Pennies*, an alternative (or additional) activity that Veronica Kell, one of the pilot teachers, found useful in judging students' comfort level with benchmark percents.

Pass the Pennies

The purpose of this activity is for students to demonstrate their understanding of benchmark percents. *Always* do this without paper and pencil. Careful observation of the members of the group will give you some valuable information. Again, if this is a struggle for your students, return to *Using Benchmarks.*

Materials: Pennies, paper, pencil

- Arrange the class in groups of multiples of 2 (4, 6, etc.). You will need 80 pennies for a group of 4, 120 pennies for a group of 6. You may play or not, based on the number of students in the class.
- Distribute the pennies such that in a group of four the distribution is 16-24-16-24.
- Tell students that there can be *no* discussion. This is a silent activity.
- Ask students to pass 25% of their pennies to the person to their right. (Students starting with 16 pennies will now have 18; students starting with 24 will now have 22.)
- Ask students to pass 100% of their pennies to the person to their right.
- Ask students to pass 50% of their pennies to the person to their right.
- Ask students to count their pennies. Each student should have 20 pennies.
- Ask students to line up their pennies, then pull out 10% (2), 15% (3), 25% (5), 40% (8), 50% (10), and 75% (15). Put the pennies back in the line between each of the "pull-outs."
- Go back and start again, with the students having their original amounts of pennies (16 or 24).
- Take out paper and pencil. Fold the paper into six columns.
- After each transfer of pennies, have the students record the number of pennies they have and the number of pennies they passed, along with the percent. They should write each of these numbers as a fraction, a decimal, and a percent. Do the same for the pull-out percents once everyone has 20 pennies.
- At the end, consider letting the students keep their 20 final pennies.

Ask students to do only a few of the problems in the *Initial Assessment.*

Making the Lesson Harder

For *Activity 1,* select publications (like *The Wall Street Journal* or a magazine on economics) that will have challenging numbers. For example, $49.3 billion, might be mentioned as an amount companies are spending on research and development.

For Task 2, ask students to do all the problems in the *Initial Assessment.*

OPENING THE UNIT IN ACTION

Two teachers who piloted Activity 1 *recounted where the open-ended search for fractions, decimals, and percents in newspapers and magazines led. The search for fractions, especially, held a few surprises.*

This activity was great—but the search for fractions created a problem. There just aren't any fractions to speak of in the newspapers these days. But my students were intent on finding them. One pair offered 3/24 as a fraction. It was from the moon-phase chart and indicated the date of the full moon (March 24). I had not anticipated this, so the discussion about doubling and halving these amounts did not work. How do you double March 24? My suggestion is to list whatever students have found on the board and discuss whether they indeed are fractions, decimals, or percents.

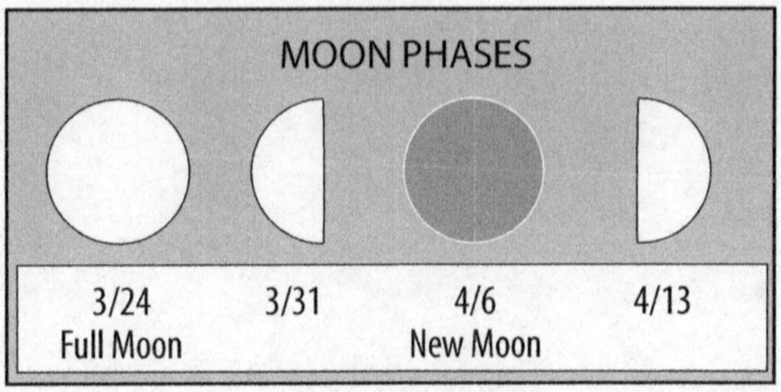

I certainly came away with two realizations: Fractions are scarce these days, and the date format looks like a fraction, even though it has nothing to do with part-whole. It surely makes you consider the concept!

Marilyn Moses
Brockton Learning Center, Brockton, MA

An interesting thing occurred in *Activity 1.* One pair of students circled 9/11 as a fraction—the date of the terrorist attack. Another pair located 24/7 (meaning "all the time") and labeled it as a fraction. But these numbers really aren't fractions. Others offered 20/20 vision and a 50/50 chance. Here the meaning is closer—a comparison of two numbers as a ratio—but still not a fractional amount you could double or halve. We had a great discussion both during the search and in the whole class's sharing. Also interesting was the question about whether 5.3% was a decimal or a percent. All in all, this was a great awareness-raising activity.

Veronica Kell
Mt. Wachusetts Community College
Devens Learning Center, Ayer, MA

FACILITATING LESSON 1

One-Tenth

How can you cut a loaf into tenths?

Synopsis

In this lesson, students explore the benchmark fraction $\frac{1}{10}$ as they visit five stations. Later they numerically describe one-tenth, including the decimal 0.1. Students brainstorm everyday names for one-tenth in preparation for the work on decimals in the next section.

1. Students compare $\frac{1}{10}$ to known benchmark fractions, using fraction strips.
2. Student pairs visit stations in which they find one-tenth of different amounts and objects.
3. Student pairs share how they found one-tenth at the different stations and discuss how they knew it was one-tenth.
4. The whole class writes one-tenth in various ways.

Objectives

- Relate $\frac{1}{10}$ and its multiples to benchmark fractions ($\frac{1}{2}$, $\frac{1}{4}$, $\frac{1}{3}$, $\frac{1}{6}$, $\frac{1}{8}$, $\frac{1}{12}$, $\frac{1}{16}$) and their equivalents
- Identify visual and numeric representations for one-tenth such as $\frac{1}{10}$, 0.1, .1, 10%
- Find one-tenth of a quantity

Materials/Prep

Set up five stations. Make duplicate stations if you have a large class. If stations are not possible, provide one station or activity per student and then rotate the station/activity. Details about station set-up and what happens are as follows:

Station 1: Paper Clips

- One box of 100 paper clips
- 100 pennies and 10 dimes (optional)

Students answer the questions: "What is a tenth of the whole box? How do you know?" Then they record their answers with their explanations (*Student Book*, p. 9).

Station 2: Your Height

- Measuring tapes (in inches)
- Various objects around the room, such as a book, a chalkboard eraser, pencils, chalk, a glass or cup, etc.

Students find objects in the room that are close to a tenth of their own heights. Then they explain their choice and reasoning (*Student Book*, p. 9).

Station 3: Index Card

- 3" x 5" index cards
- Ruler
- Scissors
- Clear tape

Each student cuts one-tenth out of an index card and tapes the tenth on the activity page (*Student Book*, p. 10). When the class shares results, you will pick a few of the tenths they cut to discuss whether they do indeed all equal one-tenth.

Station 4: One Quarter

- One quarter (25¢)
- 25 pennies for reference

Students find one-tenth of a quarter. Then they record their answers and their reasoning (*Student Book*, p. 10).

Station 5: Stamp Collection

- 4 stamps or copy and cut *Blackline Master 1: Stamps*

Students solve the following problem: "The stamps at the station represent one-tenth of someone's collection. How many stamps are in the whole collection?" They write their answers and how they arrived at them (*Student Book*, p. 11).

Have Fraction Strips for this and future lessons (see *Heads Up!* below).

Opening Discussion

Begin by saying:

In some communities, people voluntarily contribute a *tithe* for the support of their church. What is a tithe?

If nobody can define it, explain that a tithe is a voluntary donation of **one-tenth** of a person's income. Then add:

How much will someone who makes $600 per week tithe?

Ask students to think about the problem and share their thoughts with a neighbor. When all are ready, have the class share. Ask:

How did you find the tithe based on a weekly salary of $600?

Heads Up!

This is an introduction to thinking about one-tenth. Most students will likely come up with $60. Probe for why. Do not go into long explanations about how to find a tenth. Listen to students' reasoning and take notes on the board. For example, "One-tenth of 100 is 10, so $\frac{1}{10}$ of 600 is 6 x 10, or 60."

Tell students that after spending a bit of time reviewing fraction strips, they will find a tenth of various amounts as they go through the five stations.

Activity 1: Building on the Fraction Strips

Heads Up!

If students have not had exposure to fraction strips, you will need to spend some time having them fold paper strips into halves, quarters, eighths, and so forth, as described in *Using Benchmarks: Fractions and Operations*, Lesson 6. Have students practice using strips folded into halves and quarters to solve problems like $\frac{1}{2} + \frac{1}{4}$, $\frac{2}{2} + \frac{1}{4}$, or $\frac{2}{2} - \frac{1}{4}$.

The lessons in *Using Benchmarks: Fractions and Operations* gave students an opportunity to build a strong foundation of benchmark fractions using $\frac{1}{2}$, $\frac{1}{4}$, $\frac{1}{8}$, $\frac{1}{3}$, $\frac{1}{6}$, and even $\frac{1}{12}$ and $\frac{1}{16}$. This first activity will serve a dual purpose: to review those benchmark fractions and to compare the fraction $\frac{1}{10}$ and its decimal equivalent, 0.1, to the other fractions. Students will be able to compare $\frac{1}{10}$ (and 0.1) and its multiples with the benchmark fractions in order to find equivalent fractions and to estimate approximate size of tenths.

Review the term, **benchmark and the connection between division, breaking into equal parts,** and **fractions.**

Benchmarks are common points of reference used for comparison. Benchmark fractions include one-half, one-third, and one-fourth. Remember that fractions have equal parts. When we break an amount into thirds, how many equal parts do we have? In order to break an amount into eighths, how many equal parts do we have to have?

Refer students to *Activity 1: Building on the Fraction Strips (Student Book, p. 6).* Explain that there are several strips of equal size (and each strip is considered a whole). Ask them to consider what each whole amount has been equally divided into and label each fractional part.

$\frac{1}{3}$	$\frac{2}{3}$	$\frac{3}{3}$

$\frac{1}{3}$	$\frac{1}{3}$	$\frac{1}{3}$

Once students have completed *Activity 1* with a peer, ask for any "aha" moments they experienced. Record on the board and discuss.

Then explain that in future lessons they will learn how to be more exact in comparing fractions and decimals, but for now, knowing how to estimate is critical for understanding whether solutions with fractions make reasonable sense.

Activity 2: Show Me 1/10! Stations

Pairs of students visit each station to show one-tenth of various things or amounts as they answer the questions, write their explanations, and make drawings in *Activity 2: Show Me 1/10! Stations* (*Student Books*, pp. 9-11).

Station 1: Paper Clips

Students look at a total of 100 paper clips and either know that $\frac{1}{10}$ is 10 paper clips and divide 100 by 10 to find the answer, or make 10 groups to find the number in one group. In all cases, students are working with a *discrete model* of the fraction $\frac{1}{10}$.

Station 2: Your Height

Each student selects an object in the classroom that is about one-tenth of his or her height. Students estimate the length of their chosen objects. Probe for how they are making their choices. The measuring tapes available at the station are for those students who want to measure their height and/or objects.

Station 3: Index Card

Heads Up!

Try the activity at Station 3 before class, so you see the sizes of the different possible cuts and will recognize correct student answers when you see just the piece.

Students determine one-tenth of the index card, cut it out, and tape their tenths on the activity page in their *Student Books.* Note that the strip could be horizontal or vertical, and the shapes of the tenths—for example, long and skinny or short and wide—can vary. Make a note of which ones you want them to share with the class later so there is a variety to use in proving that they all represent one-tenth of the index card.

Station 4: One-Quarter

Students determine one-tenth of a quarter. For students who rely on counting or grouping, the pennies at the station will be helpful. The problem of finding a tenth of a number not evenly divisible by 10 could make this an arithmetically challenging station.

Station 5: Stamp Collection

Students are given one-tenth of a stamp collection and must determine the number of stamps in the entire collection. They have visited stations where the whole was given and they had to find the part (one-tenth); at this station, the part is given and they must find the total (the whole).

After pairs visit the stations and solve the problems, gather the whole class together. Tell them:

You have worked with one-tenth in a variety of ways. What surprised you?

Students share a few of their surprises and then review the problems at the stations. Start with the problem that involves finding one-tenth of the paper clips, and quickly establish the correct answer as 10. Move to Station 2 and examine some students' heights and the objects they chose to represent one-tenth of those heights. Ask:

How do you know this object is about one-tenth of your height?

For Station 3, gather some strips that students cut in different ways, and post them for all to see. Ask:

These strips look quite different. How could we show that these are all one-tenth of the index card?

Students might need to cut some of their tenths to "fit" them into another one, as in the following example:

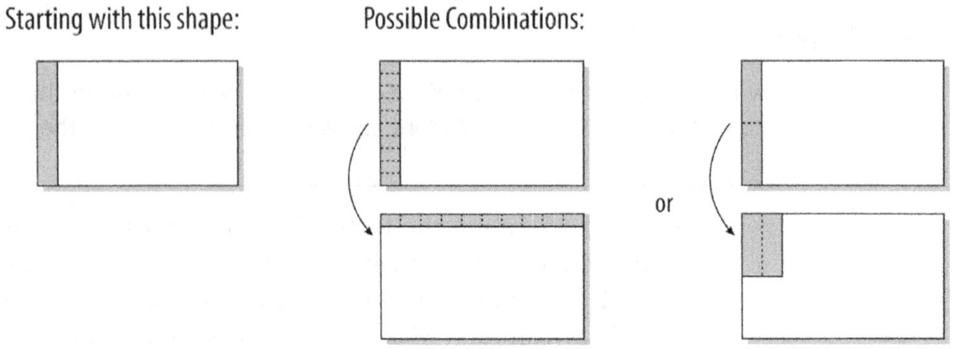

When students share their solutions for finding one-tenth of a quarter (2.5¢), probe by asking:

How do you know you have the correct answer?

Finally, Station 5 poses a different dilemma for students: They have found one-tenth of different things and/or amounts. Now they are given one-tenth of a stamp collection and asked to find the total stamps in the collection. If students find it helpful to group, they might draw ten groups, each with four stamps, to find the total. Explore their solutions further by again asking:

How do you know your answer is correct?

Direct students' attention to the Stations Wrap-Up (*Student Book*, p. 12). Give students time to write down answers individually, then write the words *one-tenth* on the board. Invite students to write responses on the board as you ask:

What is another way to write one-tenth? Who has another way?

How would your read each of these numbers?

Post the term *one-tenth* and its representations on the class vocabulary list. If all of the following notation are not mentioned, add the missing ones yourself:

- $^1/_{10}$
- 0.1
- .1
- 10%

Heads Up!

This is an ideal time to emphasize the "-th" ending as a signifier for fractions, as in one-fourth, one-fifth, and one-tenth. Some students are familiar with 10% as the equivalent to $^1/_{10}$. Probe to clarify the connection between 10% and $^1/_{10}$.

Summary Discussion

Review the ways to say, write, and describe one-tenth by posting the following chart on the board or on a piece of easel paper.

NUMBER	1/10	.1	0.1
Word Description	one out of 10 1 divided by 10 one-tenth	point one one-tenth one part of ten	zero point one one-tenth

Encourage the inclusion of several word descriptions for each column. If questions about zeroes surface, mention that there are zeroes as important and unimportant placeholders (for example, 0.1 versus .01). This idea will be examined in *Lesson 2*. Discuss the common use of "point" followed by digits—for example, saying "a three point two average." This phrasing makes it easy to avoid thinking about the meaning of the decimal point and the digits that come after it.

Ask:

What makes one-tenth a benchmark fraction or decimal?

Brainstorm some ideas about how tenths are used (to mark miles on an odometer in a car, measure medicine prescriptions, report test scores, and mark time in sports events). Also, if no one else mentions it, raise the idea of tenths being a natural extension of our base ten number system (see *Lesson Commentary* p. 18, for more on this).

Direct students to the *Vocabulary* (*Student Book*, p. 203) and *Reflections* (*Student Book*, pp. 207–208) sections before moving to the practice pages.

Practice

I Will Show You 1/10!, p. 13
For practice finding one-tenth.

Containers, p. 14
For practice marking one-tenth.

More or Less?, p. 15
For practice determining more or less than one-tenth.

Extension

More Tenths, p. 16
For using tenths to talk about halves and quarters.

Test Practice

Test Practice, p. 17

Looking Closely

Observe whether students are able to

Relate 1/10 and its multiples to benchmark fractions and their equivalents

Students may see relationships among the fractions in new ways, realizing (again) that the bigger the number in the denominator, the smaller the piece, so that tenths are smaller than eighths. Being able to visualize that 3/4 falls between seven-tenths and eight tenths should help students connect elements of prior knowledge.

Identify visual and numeric representations for one-tenth such as 1/10, 0.1, .1, 10%

Do students easily make connections among names for ¹⁄₁₀? Remind students that one-half can be written and sketched in many ways. Similarly, 10% can appear as 10 out of 100 or as one shaded row of a 10 × 10 grid, but it can also appear as one shaded square out of 10 squares. Seeing 10% should activate an association with the notion of "out of 100"—that is, the idea that percent is relative to 100 and that a fraction expressing part and whole will help make sense of the situation.

Find 10%, or 1/10, of an amount

What strategies do students have for finding one-tenth of an amount? Are students able to find one-tenth of different quantities? Note whether students have a way of finding one-tenth that they use consistently or whether they change their strategy given the situation. For example, they might sort objects into ten piles, divide by ten, move the decimal point, or count out groups of 10 and then select one of the set. In each case, ask them to explain how they know they have determined one-tenth of an amount.

Work with them to find 10% in one or more of these ways:

- Break a number into smaller parts, find $\frac{1}{10}$ of each, and then combine the results. For instance, to determine $\frac{1}{10}$ of 200, find $\frac{1}{10}$ of 100 (10), and then double that amount.
- For small, familiar numbers, such as 20 or 40, imagine sharing an amount equally among 10 people, or, if necessary, draw a picture to show an amount divided into 10 equal parts.
- Approximate to a multiple of 10 or 100, and then take 1/10. For instance, $\frac{1}{10}$ of 153 is about 15, because 153 is close to 150 and $\frac{1}{10}$ of 150 is 15.
- Think of an everyday situation where you would find 10% (such as calculating a tip or tax). For instance, to find 10% of $153, you could just use

what you know about place value and move the decimal point to the left one place to get a number one-tenth the amount—$15.30, in this case. Or you could think in terms of 10¢ for each dollar.

- If the original amount is less than one, e.g., a quarter of a dollar or a quarter of a cup, remind students that a tenth is one of 10 equal parts. Start with the situation where the sub-amounts are easier to quantify. Twenty-five cents in 10 equal parts is 2.5¢. Thinking about a fraction to represent 2.5¢ of a whole could be challenging. Students will have to imagine 4 quarters, each divided into 10 equal parts, so 40 parts, each worth 2.5¢. Can they apply this kind of thinking to finding 1⁄10 of a quarter of a parking lot with 100 spaces?

Introduce a number line to reinforce the fraction-percent relationship. Label the line using percents and fractions, as shown:

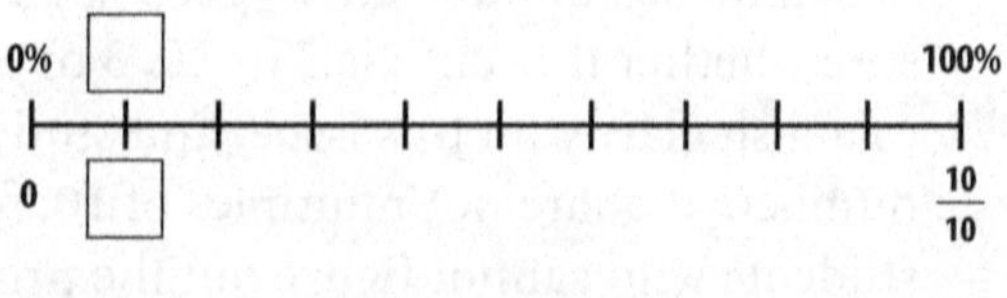

Thinking about cents per dollar is another method for making the connection between the fractions and percents, but this becomes less helpful as numbers get larger. Division by 10 is more applicable for such numbers. This method needs to be solidly understood for students to be successful in the activities in *Lesson 2.*

Rationale

This lesson lays the foundation for thinking about percents as fractions and as relationships defined by the total 100. It might be tempting to think of percents as whole numbers because they can be added in the same way as whole numbers. However, thinking of 10% as the fraction $\frac{1}{10}$, or $\frac{10}{100}$, keeps the meaning of the relationship between part and whole prominent.

Math Background

Finding one-tenth of 10, 20, 30, 40, or any multiple of 10 is relatively simple. When those amounts are larger—say 200, 300, or 400—students are not always sure whether the tenth is 2 or 20, 3 or 30, or 4 or 40. The activities in this lesson establish clarity on this issue, and students can move on to finding one-tenth of numbers that are not multiples of 10, for example, 25. By using manipulatives, students who cannot figure out the problem mathematically are able to arrive at the solution. Alternatively, given a number or quantity that represents one-tenth of a total, how can the total be found? This problem forces students to break the whole to make 10 parts and—in the case of finding the whole—put the parts together again to make the whole. Students will have worked with these same ideas with whole numbers in the *Everyday Numbers* unit, and they now extend these concepts to fractions, decimals, and percents.

Context

In the United States we use the English Standard system of measurement—inches, feet, and yards. However, there has long been a movement toward converting to the metric system based on the number 10. Widely used around the world, the metric system makes it easy to compute. Regardless of the standard of measurement used, one-tenth is a benchmark fraction, decimal, and percent. In everyday life, our monetary system is based on 10; for example, one dollar has 10 dimes and 100 pennies. Decimals are used extensively to describe measurements, time, and money (e.g., 1.2 million dollars). Percents, also based on 10, are widely used to describe information (e.g., 10% increase, 6% tax, 55% of voters, etc.).

Facilitation

Most of this lesson is focused on reviewing the findings at the stations students visit, with the emphasis on finding a tenth and representing it in various ways. The stations provide situations that help students understand and problem-solve ideas about tenths. Do not push for formulas, rules, or tricks you know. If students bring them up, ask for an explanation of their meanings and how they work.

Making the Lesson Easier

If you see that students are struggling at any particular station, encourage them to use pennies, paper clips, or any other easy-to-count objects to make sense of the problem. Then if you see them struggling again at another station, ask them how what they learned when they used the manipulatives could help them in this new situation.

Making the Lesson Harder

If the numbers at the stations are too easy for some students, use less "friendly" numbers, but only after you have ascertained that these students have a true visual understanding. Ask students to use a picture or drawing to show their reasoning and to make up a problem involving tenths for others to solve after first solving it themselves with a picture.

FACILITATING LESSON 2

More About One-Tenth

Where is the tenth?

Synopsis

In this lesson, students continue working with the decimal form of one-tenth (0.1) and the percent form (10%) to understand their relationship to the fraction 1/10.

1. Students explicitly examine and share what they already understand about place value.
2. Students respond to three "hard questions" about place value.
3. Students focus on how 10% is 1/10.
4. The whole class reviews names and descriptions of one-tenth.

Objectives

- Identify representations equivalent to tenths
- Examine and specify the role of place and the decimal point in a number's value

Materials/Prep

- Calculators for *Math Inspection*
- *Blackline Master 2: Match Cards*

Copy and cut sufficient copies of *Match Cards* so that each student receives at least one card at the beginning of *Activity 1: One-Tenth Match* (*Student Book*, p. 20), and then receives a complete, uncut copy of the *Match Cards* sheet at the end.

Opening Discussion

Review quickly the *Summary Discussion* at the end of *Lesson 1*, p. 15. Say:

We said there were many ways to represent one-tenth. Remind me of some of them.

As students share, make sure that all the fraction, decimal, and percent names for one-tenth are listed. They will need to recognize them in *Activity 1: One-Tenth Match* (*Student Book*, p. 20).

Activity 1: One-Tenth Match

Distribute to each student one or more of the *Match Cards*. Instruct students to think of as many ways to represent the value on their cards as possible, then mingle until they find their card's match in words or numbers. Note that there are multiple matches for some of the cards, but they need only find one card to match each of theirs at this point.

Every pair of students posts their matching cards so everyone can see them. Students then review the pairings and post all the cards they think represent one-tenth (*Student Book*, p. 20).

Ask students:

If you have a card that shows one-tenth, stand by your card. Does everyone agree that all these cards are equivalent? Why or why not?

Post all the one-tenth cards together, and distribute copies of the full, uncut *Match Cards* sheet so students can check off the matches equal to one-tenth.

Are there any other cards that show one-tenth that we didn't mark off?

Invite students to share their ideas and images for one-tenth. Lay the groundwork for using one-tenth as a benchmark for comparison.

When you think about one-tenth, do you picture it in one of these ways? If not, how do you picture it?

Of the cards that are left, how would you organize them from smallest amount to largest? Why?

How does one-tenth compare with the benchmark fractions? (Smaller than ¾, ½, and ¼)

Activity 2: Hard Questions

Heads Up!

Did you know that the convention of using a point as way to separate the whole number from the fraction differs from country to country? For example, some countries use a comma for a decimal marker, and $1,5 million means $1½ million. Are there people from other countries in your class? Have a discussion about the differences they have noticed. You might want to read some background comparing different countries' conventions at http://robertspage.com/cultr14.html and *The Answer is Still the Same* at https://www.terc.edu/adultnumeracycenter/wp-content/uploads/sites/28/2025/09/The_Answer_Is_Still_the_Same_2020.pdf.

This activity is an opportunity for students to share with one another what they already understand about place value. Listen carefully, verifying correct conceptions, and questioning misconceptions that might arise.

Ask a volunteer to read the cartoon and paragraph aloud at the beginning of *Activity 2: Hard Questions*, (*Student Book*, p. 21). When everyone is clear about the directions, encourage students to talk in pairs about an approach to helping each child. Ask each pair to record their parent responses.

Homework Problem 1

When every pair has recorded at least one possible parental response to each child, pull the class together. Write the youngest child's homework problem on the board. Ask:

Who wants to share how they responded to the youngest child?

Did anyone say something different?

Are there any ideas that you would like to question?

Listen carefully to what students suggest, and record all ideas on the board. For example, the following might be ideas that students bring up:

Circle the larger number: 25 52

Young child's question: Daddy, isn't two five the same as five two?

Father's response: I would tell her that 25 doesn't mean two five. It means twenty-five. And 52 means fifty-two.

Maybe she needs practice counting. I would tell the father to get out a jar of pennies or a bag of peanuts, and have her count out twenty-five in a pile. And then count out fifty-two in another pile. Then I would ask her: Does that look the same? Which one looks like more? Then, maybe have her make small piles of 10 so she could see 25 as 2 tens and 5 ones, and 52 as 5 tens and 2 ones.

Summarize, making sure that the idea of differentiating between the tens and ones places has been a focus. If not, explicitly address it, and probe further with these questions:

- In 25 and 52, the first digit tells how many tens and the second digit tells, how many ones. The 2's and 5's have different values because of where they are.
- What about 09 and 90?
- What about a 3-digit number? 4-digit? 5? 6? 7?
- Does anyone know how to complete a place-value chart? (Set up a simple place-value chart on the board.)

______	______	______	______	______	______	tens	ones

- What pattern do you see? (Each place has a value ten times greater than the place to its right, or one tenth of the value of the place to its left.)
- How would you write 870,200 in the place-value chart?

Homework Problem 2

Now write the middle school child's homework problem on the board. Ask:

Who wants to share how they responded to the middle school child?

Did anyone say something different?

Are there any ideas that you would like to question?

Listen carefully to what students suggest, and record all ideas on the board. For example, the following might be ideas that students bring up:

> Compare these kids' temperatures: 97.9° 99.7°
>
> Middle school child's question: Doesn't 97.9 mean the same as 99.7?
>
> Father's response: I would tell her that 97.9 means 97 degrees and nine-tenths of a degree. 99.7 is almost 2 degrees warmer.
>
> Maybe she needs practice "seeing" tenths. I would tell the father to draw a number line from 90 to 100 and mark those temperature readings. If he had a thermometer, he could even take her temperature and put that on the number line, too. Then he could ask: Which one looks like more? How much more?

Summarize, making sure that the ideas of differentiating ones and tenths places and of a decimal point as a separator have been a focus. If not, explicitly address them and probe further, with questions such as these:

- What does a point tell you?
- Where would you place 97.9 on the place-value chart? What about 99.7?

______	______	______	______	______	tens	ones

Homework Problem 3

Heads Up!

This is a key moment. The relationships of multiples of 10's are easily lost when the numbers become unwieldy. Most of us don't regularly have to work with quantities that lead us to feel solid on these relationships, e.g., the idea that 10,000 one hundred times is a million. Take some time to allow students to observe and absorb patterns.

Finally, write the teenager's homework problem on the board. Ask:

Who wants to share how they responded to the teenager?

Did anyone say something different?

Are there any ideas that you would like to question?

Listen carefully to what students suggest, and record all ideas on the board. For example, the following might be ideas that students bring up:

> A newspaper article: Yesterday, the city threatened to slash the school budget by $750,000. Today, the actual vote was to cut by only $.5 million.
>
> Teenager's question: Why does it say "only"? A million is more than a thousand.
>
> Parent's response: I would tell her that yes, 1 million is more than 1,000, but you don't have a whole one million. A half of one million is .5 million, which is the same as 500 thousand. And 500,000 is less than 750,000.

Summarize to return the focus to the skill of expressing large numbers with decimals. If not, explicitly address that idea, and probe further with questions such as these:

- Newspapers often express large amounts with decimals. For example: a news reporter says "campaign spending exceeds $1.7 billion." How would you write that out in numerals?

  Write 1 thousand, 10 thousand, 100 thousand, 1 million, 1 billion, and 1 trillion as numerals. Compare the size of the numbers. What pattern do you notice?

  [A million is a thousand thousands. 1,000,000 = 1,000 thousands.
  A billion is a thousand millions. 1,000,000,000 = 1,000 millions.
  A trillion is a thousand billions. 1,000,000,000,000 = 1,000 billions.]

- Which is greater? Explain your thinking.

  $25 million -or- $.1 billion?

  $1.5 billion -or- $150 million?

Heads Up!

A billion in the U.S. is 1,000 millions. A billion in the United Kingdom at one time meant 1,000,000 millions. However, most people now use a billion to mean a thousand million. See oxforddictionary.com for more detail.

Suggestion: Ask students to look around and bring in examples in the newspaper headlines—or online—for large amounts (in the millions, billions, or trillions) expressed with decimals. You might then have students work with the numbers; for example, ask them to convert those decimals to numerals and rewrite the sentences, or evaluate the story based on those numerical representations.

Activity 3: Why is 10% Equal to 1/10?

Refer students to *Activity 3* (*Student Book*, p. 23), allowing time for each person to show why 10% equals $\frac{1}{10}$. Students need to see 10% as ten out of one hundred. This activity reinforces students' understanding of the fact that the 10 in 10% is not related to the 10 in $\frac{1}{10}$. The relationship that should be at the forefront is that 1 out of 10 has the same value as $\frac{10}{100}$. Emphasizing this distinction is important so that students don't generalize and think of 20% as $\frac{1}{20}$ rather than as $\frac{20}{100}$ or $\frac{1}{5}$.

Math Inspection: Entering Numbers into a Machine

This inspection is meant to help students understand and be able to show what happens when a number is entered into a calculator—why is there a decimal at the end, and what does it mean? It allows students to pay attention to place value as they enter numbers with decimals, usually money, and read them as dollars and cents.

Math inspection pages will have more impact if students are given time to notice patterns, come to generalizations, and justify their reasoning. Ask students to enter numbers, look for patterns, and make a statement about what the display looks like and what it means.

Summary Discussion

Review the ways to say, write, and describe one-tenth by posting the foillowing chart on the board or digitally.

Number	1/10	.1	0.1
Word Description			
Graphic			

Encourage the inclusion of several word descriptions for each column and a different graphic for each of the tenth numbers (grids, diagrams, etc.).

Articulate observations about zeroes, including trailing zeroes and zeroes as important and unimportant placeholders (for example 0.01, versus .1). Ask:

One tenth is a benchmark fraction, decimal, and percent. Where do you see one-tenth (1/10, 0.1, .1, and 10%) used in your daily life?

On the class vocabulary list, ask students to explain the terms "0.1 and .1," and create a working definition for *decimal point.* Close by inviting students to write about what 0.1 means to them and how they find 0.1 of a quantity in *Reflections* (*Student Book*, pp. 208-209).

Practice

Make a Gauge, p. 26
For practice marking increments of one-tenth and labeling them as decimals.

Location, Location, Location, p. 27
For practice locating decimals on a number line-like map and adding decimals.

Gaining Weight, p. 28
For practice finding 0.1 of a quantity.

Digits, Places, and Points, p. 29
For place value practice as a follow-up to *Activity 2.*

Grid Visions, p. 33
For practice finding 0.1 of related amounts.

Test Practice

Test Practice, p. 35

Looking Closely

Observe whether students are able to

Identify multiple representations equivalent to one-tenth, such as 1/10, 0.1, .1, 10%, and visual models

Are students able to match up the cards that represent one-tenth in a variety of ways? Check students' work when they review matched cards and list those equal to one-tenth to see whether they are able to identify the various representations of one-tenth. Where there is disagreement, encourage discussion, asking students to prove their points. Ask *how* they know a card is or is not equal to one-tenth.

The conversation about those cards *not* equivalent to one-tenth is also important. Do students recognize the difference between .1 and 1.? "Less than a whole" and "more than a whole" are important ideas, as is the whole, 100%, and the entire shaded grid.

Examine and specify the role of place and the decimal point in a number's value

Are students paying attention to the decimal point and numbers around it? Do they use contextual clues to confirm their instincts or do they try to gloss over the decimal? Do they see how the decimal sections off partial and whole amounts within the number and how partial amounts may be but are not always less than one (as in 7.5 million)? Pay attention to comments that relate these ideas to prior knowledge, e.g., from measuring liquids or weighing items.

Rationale

Tenths are critical to an understanding of money, measurement, percents, and scientific notation. Because decimals look much like whole numbers, students often complete operations with them without solidly understanding their meaning and, therefore, misinterpret solutions. They might multiply 0.2 x 5 and write "10" rather than "1." Clarifying the fraction-decimal relationship solidifies students' understandings of the decimal values and aids their reasoning about solution values.

Associating the decimal fraction with the common fraction is as much a language issue as a mathematical issue. Help students establish the connection between the two by using the term "one-tenth" to refer to both decimal and common forms of the fraction. Practice is key to forming solid associations.

Math Background

We use a base 10 number system. Counting to 10, counting by 10's, the fraction one-tenth, the decimal 0.1, and the percent 10% are all part of our everyday usage of number. However, it is also the case that students are often not sufficiently grounded in their understanding of all these forms to be able to manipulate them correctly later when they operate with tenths. One-tenth of a 10-piece candy bar is easy to find; it is not as simple to figure out one-tenth of a 12-piece candy bar. Understanding tenths, then, often involves understanding parts and wholes and the relationship between them.

Historical Background: The Hindu place-value system was in use for nearly 1,000 years before decimal fractions were introduced in the 16th century as a way to notate square roots of irrational numbers. The decimal point was not commonly used to separate whole numbers and decimal fractions until the first half of the 1700's. There is still no universal way to represent decimal fractions. Some people use a point to separate whole numbers from decimals, as we do in the U.S., while others use a comma, as the British do: 53,25 vs. 53.25. (See *Historical Topics for the Mathematics Classroom*, NCTM, 1989, for more.)

Context

Punching amounts into an ATM machines is an example in which the decimal is static and numbers change in value around it, depending on their position. If students are familiar with this context, ask them to picture themselves entering amounts and seeing them displayed on the screen as the numbers increase from pennies to dimes to dollars to amounts in the 10's and 100's of dollars.

Facilitation

Although the ideas and activities may seem trivial at first glance, our field-test experience tells us that students' conceptions of tenths in their various forms (fractions, decimals, and percents) are not solid. Urge students to demonstrate their understanding, to share it with you and their peers, and to think about the different forms of notation.

Making the Lesson Easier

For students struggling to match the cards, start them with fewer cards and omit those that are greater than one. For students familiar with circle graphs, share how you would illustrate the first example of *Activity 3* using a circle. Stress that the circle represents the whole and the two parts, though not precisely accurate, should show $1/10$ and $9/10$.

Making the Lesson Harder

Connect decimal and whole number place value with scientific notation (powers of 10). The book *Powers of Ten*, by Philip and Phylis Morrison, is an excellent visual and mathematical resource on the subject.

FACILITATING LESSON 3

What Is Your Plan?

How can you use percents to plan the layout of the space?

Synopsis

In this lesson, students build on their knowledge of 10%, or 1/10, to calculate percents that are multiples of 10%. The focus is on an area model of percents, with students repeatedly using grids to demonstrate percents for display spaces.

1. The whole class discusses strategies for finding 20% of a given amount and considers the applicability of these strategies to other multiples of 10%.
2. Individually or in pairs, students allocate display space in a market stall to practice finding 10%, 20%, 30%, and 40% of a total of 50 square feet.
3. In pairs, students decide how to allocate space for another market stall, determining the percent they will assign to each item so that the entire space (100%) is used.
4. Students work with a partner to consider the fraction equivalents of 20% and then name equivalent fractions for 10% to 100% and 5%.
5. Given 1/10, students determine the whole and prove their findings with grids.
6. Students record percent, decimal, and fraction equivalents and summarize their learning.

Objectives

- Share strategies to determine multiples of 10% of a given amount
- Determine a total, given 10% of it
- Base decisions involving percents on the fact that the whole of an amount or space equals 100%

Materials/Prep

- Calculators
- Sticky notes
- Tape
- Copies of building blueprints (optional)
- Colored pencils/markers in at least three colors, for student use
- *Blackline Master 11: 50-Block Grids*, one to two per student

For *Activity 3:*

- Post a large version of *Blackline Master 3: Two Out of Twenty*
- Prepare to display a 5 × 2 grid with the following key:

My Brother's Display Plan

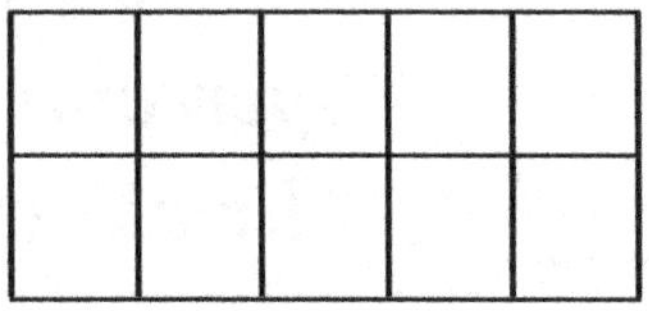

Key to Display Items on the Plan
A—stuffed animals (50%)
B—silk flowers (40%)
C—perfume (10%)

For *Activity 4*:

- Collect a large quantity of small objects, e.g., beads, tiles, toothpicks, or paper clips
- Prepare to display the following chart during *Activity 4*:

$\frac{1}{10}$, or 10%, of Material for the Design	Rule for Finding the Whole When You Know $\frac{1}{10}$, or 10%	Design Total = $\frac{10}{10}$, or 100%
1		
2		
3		
4		

Opening Discussion

Remind students of their work with 10% and 1⁄10 and then ask:

What if you saw a sign for a 20%-off sale? How would you find 20% of an amount?

Record the strategies mentioned by students. Make sure the list includes the following methods:

- Finding 1⁄10 (dividing by 10) and multiplying by two (or doubling);
- Multiplying by 0.2, or 2⁄10, on the calculator.

Ask students:

Will these same strategies work to find 30% of an amount? How do you know?

How could you figure out the cost of items at a 70%-off sale?

It is tempting for most people to follow the same pattern and take 70% off the cost and then subtract that sale amount from the original cost. Some students may suggest finding 30%, which will be the amount you will pay for the item.

Heads Up!

Often students think of a percent in terms of *x* cents to the dollar, which they add up to determine the percent of a whole. This strategy works well with smaller numbers but is less efficient with larger numbers, as students will discover. Sharing other approaches provides alternative methods for them to consider.

Allow time for students to practice a few examples and to explain their solutions. They might choose to work with 50% values in combination with 20% values.

Tell the class that they will find some or all of these strategies helpful as they create display plans for market stalls in *Activity 3.*

Activity 1: 20% Dilemma

Tell the class that you overheard two people arguing about how to find 20% off.

One person said she could find 20% by dividing the total amount by 20. The other person said that method would not work at all.

Choose a partner and take a minute to think this through: If 10% is $\frac{1}{10}$, is 20% one-twentieth?

How do you know?

Refer students to *Activity 1: 20% Dilemma* (*Student Book*, p. 38) and suggest they use the grid to show their reasoning.

The answer $\frac{20}{100}$ is acceptable, as is $\frac{2}{10}$, but keep looking for the equivalent that puts *one* in the numerator ($\frac{1}{5}$). Listen to students' reasoning, and connect their shaded grids to the various fractions presented. If no one has shaded a grid in a manner that makes one-fifth visible, provide one (see *Looking Closely*, p. 40, for examples).

What percent is $\frac{1}{20}$? (5%)

What fraction is 20%? ($\frac{1}{5}$)

If 20% is $\frac{1}{5}$, what percent would equal $\frac{2}{5}$? (40%) Explain. (Possible explanation: $\frac{1}{5}$ is $\frac{2}{10}$, which is $\frac{20}{100}$, or 20%. $\frac{2}{5}$ is $\frac{4}{10}$ which is $\frac{40}{100}$, or 40%.)

Now cement the generalization by saying:

So any percent can be written as a fraction with the part over the whole.

If 10% equals 1/10, what does 30% equal? (3/10) What does 70% equal? How do you know?

Name each of the percents mentioned during *Opening Discussion* (20%, 30%, 70%, 50%), as a fraction with a denominator of 10, and post these fractions with their percents. Name percents as fractions with 100 as a denominator to reinforce the association of percents with parts per 100, but do not overload the chart with information.

Also include the fraction 1/5 and its multiples and the fraction and percent equivalents for 1/20 on the class table.

Squares Representing Common Percents and Fraction Equivalents										
5% 1/20	10% 1/10	20% 20/100 2/10 1/5	30% 3/10	40% 4/10 2/5	50% 50/100 5/10 1/2	60% 6/10 3/5	70% 7/10	80% 8/10 4/5	90% 9/10	100% 10/10 1

Activity 2: Display Plans

This activity has two parts. First students design the layout of products for a fresh fruit market stall. The four categories of fruit take up 40%, 30%, 20%, and 10% of the space. During the second part of the activity, students design their own market stalls and determine the percentage of the entire display area used for each product.

Model the design of the display space using your copy of "My Brother's Display Plan."

My brother sells gifts at the mall. He drew up this plan because he wanted to arrange his merchandise in a new way. The display space in his stall is five feet across and two feet wide, or 10 square feet.

He uses 50% of the space for stuffed animals, 40% for silk flowers, and 10% for perfume in bottles, as described in the key.

What percent does each square represent? How do you know?

When he uses *all* of his space, what percent is covered? How do you know?

Ask students to label (or color in) the squares to show one possible arrangement for the given percents.

Either of these arrangements would work with the given percents:

A	A	A	A	A
F	F	F	F	P

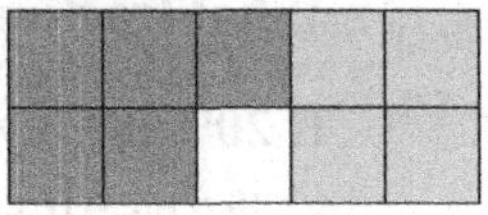

Ask for two or three volunteers to share their arrangements for the market stall. Verify that each display plan matches the market stall percents, asking, for example:

How did you decide to use five squares for animals, one for perfume, and four for flowers? What makes you certain these numbers are correct?

Point out that even though arrangements may differ, each should show the same percent of stuffed animals, silk flowers, and perfume.

Heads Up!

Highlight the relationship between 10% and 40% or 50%, so students see that by finding 10%, they can then easily find any multiple of 10%.

Refer students to *Activity 2: Display Plans* (*Student Book*, p. 39), and ask them to work individually or in pairs. Distribute *Blackline Master 11: 50-Block Grids* to each student.

As students finish, they post their plans and review what others have posted. They use sticky notes to mark plans about which they have questions.

When everyone is done, call the group together. As time permits, raise some or all of the following questions:

What are some ways to calculate 20% of the 50-square grid?

Students may find that 10% is 5 and double the 5 to 10 to get 20%. They may know that 20% is $\frac{1}{5}$ and therefore divide 50 by 5 to get 10. They could multiply 50 by .2 ($\frac{2}{10}$) using a calculator to get 10.

What are some ways to calculate 30% of 50?

What fraction does 30% represent in this grid? (What is the whole and what is the part?) ($\frac{15}{50}$ or $\frac{3}{10}$ depending on labeling)

Did you ever use a certain percent to find another—20% to find 40%, for instance? How?

Did anyone make a display plan that used too many or too few squares by mistake? How did you find the problem? How did you fix it?

Some students may realize that each square represents 2%. Others might find 10% of the total number of grid squares and use multiples of 10 to find the total number of squares for other percents. Discuss ways to check answers: The number of 20% and 30% squares must equal the number of 10% and 40% squares, for instance; and the numbers of squares for multiples of 10% must relate to the multiple itself.

Practice finding multiples of 10% of a new number, such as 200, to solidify the use of 10% as a benchmark.

Activity 3: What Is *Your* Plan?

Direct students to *Activity* 3, (*Student Book*, p. 41). Students will use the other 50-block grid from *Blackline Master 11.* Ask students to devise a plan for the cart type they choose.

Before student pairs begin to color their grids, check to see whether the percent of space they have allocated equals 100%. Ask:

How can you figure out whether you have used the entire display space?

When pairs finish, ask them to exchange designs to check each other's work. Summarize by asking:

What was the total space you used as a percent? (100%) **What was the total space in square feet?** (50 square feet)

What percent of the space was used by one of your items? What fraction is that? (What is the whole; what is the part?)

How did you figure out how many squares that percent equaled?

Activity 4: Here Is 10%; What Is the Whole?

Direct students to *Activity 4 (Student Book*, p. 43.) Give different numbers of small objects (between two and 10) such as beads, tiles, paper clips, or toothpicks to each student. Set the stage for the activity by saying:

Each of you has $\frac{1}{10}$ of the total material you need to make a design. What is the total number of objects needed?

If students have difficulty, suggest they think of the fraction one-tenth as 1 part of a whole made of 10 parts, or as 1 out of 10.

If one bead represents $\frac{1}{10}$ of the total material needed, what would the fraction name for the total be? ($\frac{10}{10}$) **How do you know?**

How many times bigger is the whole than the part? (10 times bigger)

If students start with two beads, this number will grow 10 times. Drawing a rectangle and shading parts might help them see the part and the whole.

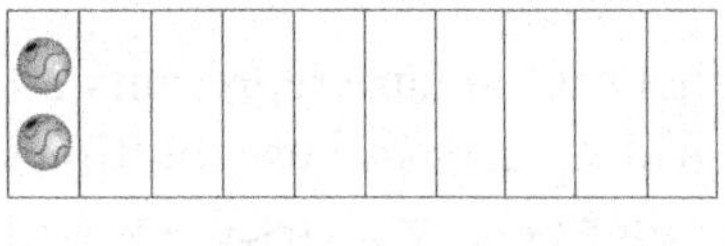

What percent could you use to describe your amount?

Another way to picture this is to draw a circle of 20 beads. Draw lines between beads to indicate a 10% amount.

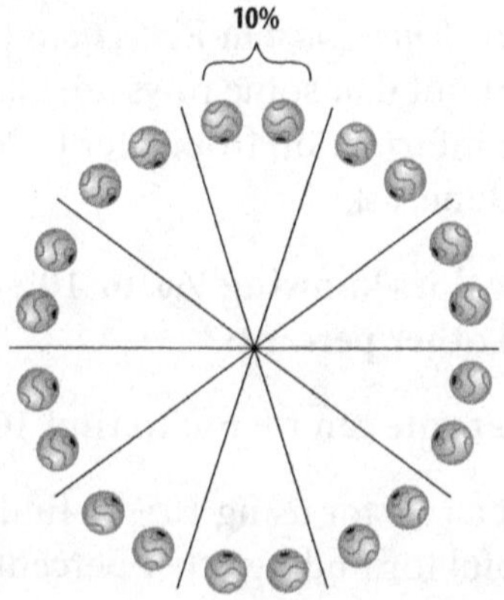

Give students time to complete the first question of *Activity 4*. Those who finish early can work on the last two questions. Bring the class back together and ask them to share their results, beginning with the person who has just two small objects.

Organize students' findings in the chart. Begin with the person who has just two beads or tiles. Then take responses from others. Your table should look something like this:

$\frac{1}{10}$, or 10%, of Material for the Design	Rule for Finding the Whole When You Know $\frac{1}{10}$, or 10%	Design Total = $\frac{10}{10}$, or 100%
1	×10	10
2	×10	20
3	×10	30
	Rule: Multiply by 10	

Ask:

What rule or pattern describes a way to find the whole, or 100%, when you know 1⁄10, or 10%, of it?

Students should explain that you can multiply the value of 1⁄10 of the total by 10 to find the total.

After completing several rows, explore with students the reasoning behind the rule to demonstrate how the rule can be generalized.

Why does this rule work?

What if your beads represented one-fourth of the whole? How would you find the whole then?

Summary Discussion

Invite students to complete the *Fraction/Decimal/Percents Equivalents Chart* in *Reflections* (*Student Book*, p. 210). Point out that some rows are blank but will be filled in with information from later lessons. When most students are done, ask:

How does knowing $\frac{1}{10}$, or 10%, of an amount help you find other percents?

What rule can we use to find 10% and then calculate other percents?

Post a rule for using 10% to find other percents and reasons why finding 10% is helpful for finding other percents.

Direct students to *Vocabulary* (*Student Book*, p. 204), to describe multiples of 10% and discuss any additional terms they would like to add to the list.

Then ask each student to say one thing they plan to remember from this lesson. Summarize any commonalities expressed.

Practice

Controlling Costs for Seniors, p. 45
For practice estimating 10% of three-digit numbers.

Money Down, p. 46
For practice finding the whole.

Drugstore Markups and Markdowns, p. 47
For practice finding 10% off and a 10% markup.

More Plans, p. 48
For practice creating a display plan on an L-shaped grid.

Visual Percents, p. 50
Provides practice matching percents with circle graph pictures, marking graph increments, showing the whole when you know the part, marking number lines, and shading grids.

Comparing Percents, p. 55
Students use the symbols <, >, and =.

Extension

Rami's Interest, p. 57
For practice calculating percents, using 10% and 20%.

Test Practice

Test Practice, p. 59

Looking Closely

Observe whether students are able to

Share strategies to determine multiples of 10% of a given amount

Can students find a multiple of 10%?

Work with students to use one of these approaches:

- *Use $\frac{1}{10}$.* Ten percent is $\frac{1}{10}$ of a number, so 20% is twice $\frac{1}{10}$, 30% is three times $\frac{1}{10}$, and 40% is four times $\frac{1}{10}$. Emphasize the multiplicative relationship by asking questions such as "What do you multiply 10 by to get 20?"
- *Use a percent to find another.* Although many students will find 10% (or $\frac{1}{10}$) a good starting point for finding other percents, they can try using percents other than 10% as well. For instance, if they know 50% of a total amount, they can divide that amount by five to find 10%. If they know what 20% is, they can double that to find 40%.
- *Convert to a familiar fraction.* Students with a good grasp of fractions may find that sometimes it is easiest to convert a percent to a familiar fraction: 30% of 50 is the same as $\frac{3}{10}$ of 50; 20% of 50 is the same as $\frac{2}{10}$, or $\frac{1}{5}$, of 50.
- *Subtract from the whole.* Some students will use the whole to find the part in cases such as the following: If 20% is 10 squares and 50% is 25 squares, the number of squares in the remaining 30% can be found by subtracting the number of squares in 70% (25 + 10) from the whole: 50 – 35 = 15 squares, or 30%.

Though many adults use a shortcut to find 10%, moving the decimal point, the reason why this works is not always clear to them. Connect the idea of dividing by 10 (finding $\frac{1}{10}$) to place value (moving the decimal one place to the left). Having found 10%, they can double, triple, or quadruple the 10% amount accordingly.

Determine a total, given 10% of it

How do students think about the task? They may find it helpful to think about the 10% amount as $\frac{1}{10}$. They can repeatedly add $\frac{1}{10}$ to $\frac{1}{10}$ until they find the whole, $\frac{10}{10}$. Or they may double the percent amount, working up by increments to 100%, the whole. The chart created during Activity 4 (*Student Book*, p. 43) should help everyone see that in a situation where 10% is given, multiply by 10 to find the whole. This is another opportunity to point out the inverse relationship of multiplication and division. Dividing the whole by 10 yields 10%, or $\frac{1}{10}$; multiplying $\frac{1}{10}$ by 10 yields the whole.

Students who have studied percents before may want to use a proportion algorithm, e.g., $\frac{2}{x} = \frac{10}{100}$, or $\frac{1}{10}$. The temptation to cross-multiply to arrive at an answer might eclipse the opportunity to think through the meaning of the situation. Ask students who are attached to one way of finding a solution to prove their answers by using a different method.

Determine for a variety of part-whole situations whether each shows an amount more than, less than, or equal to 10%.

Do students have strategies for determining whether an amount shown in a part-whole situation is equal to 10%? Ask students to explain what they see as the whole, what they see as the part, and how that part amount compares with 10%, or $\frac{1}{10}$. Make sure that they see 10% and $\frac{1}{10}$ as two names for the same amount. Work with students on different methods for finding 10%. By shading grids, marking number lines and partitioning objects or pictures, students can learn to see the fraction-percent relationship.

Remind students that one half can be written and sketched in many ways. Similarly, 10% can appear as 10 out of 100 or as one shaded row of a 10 x 10 grid, but it can also appear as one shaded square out of 10 squares. Seeing 10% should trigger for students the idea that percent is relative to 100 and that a fraction expressing part and whole will help make sense of the situation.

Base decisions involving percents on the fact that the *whole* of an amount or space equals 100%

Do students understand that the whole of anything equals 100%? Work with students to count up by 10% until they reach 100%—10%, 20%, 30%, and so on. Record the term "10%" each time they count a multiple of 10, and then ask them to total the figures.

You might also start with complements by asking questions such as "Ninety percent of the work is done; what percent is left to do?" Create a list of complements as pictured here.

90%	10%	100%
80%	20%	100%
70%	30%	100%
60%	40%	100%
50%	50%	100%

Next work with sets of three percents. Give students two of the percents, and ask what the third is. As they become more comfortable, provide one percent and ask what the two others *might* be. Use visual examples with grids as well. Emphasize that 100% is the same as $\frac{100}{100}$. This concept will continue to be reinforced throughout the unit.

Rationale

Knowing how to find 10% of an amount unlocks the door to finding any percent of any amount. Noticing patterns to make generalizations builds on students' number sense and gives them a way to simplify the task of finding percents.

Math Background

Numerate adults often use 10% as a benchmark for determining other percents. Seeing percents as they relate to whole numbers is helpful when finding percents—for example, if you double 10%, you get 20%; if you triple 10%, you get 30%. The "rules" about moving the decimal point are only helpful when the concept of finding 10% has been mastered so as not to cause confusion about whether the point is moved to the left or right.

To find a percent of a *total* amount is a different problem. If the total amount is 100, the task is quite simple; e.g., 20% of 100 is 20, and 50% of 100 is 50. But when the total is a different amount, say 50, the problem is more challenging. To find 30% of 50, it is helpful to know that 10% of 50 is 5; since 30% is triple 10%, then 30% of 50 is 3 x 5, or 15. The reasoning here is important, even if multiplication is used later on (30 x 50 =150), so the decimal point can be placed correctly.

Percents are also commonly seen as fractions (e.g., 10% = $\frac{10}{100}$, or $\frac{1}{10}$). Two misconceptions surface when working with percents as fractions: first, that 20% equals $\frac{1}{20}$, not $\frac{1}{5}$; and second, that to find $\frac{1}{20}$, you find $\frac{1}{10}$ of $\frac{1}{10}$ (perhaps because to find $\frac{1}{4}$, you find $\frac{1}{2}$ of $\frac{1}{2}$). These can easily be corrected by using graphic representations.

To reinforce understanding of the multiplication or division operations used to find percent of a total, visual representations are helpful. When shading 10 squares in a 50-square grid, it helps to visualize that if the grid had 100 squares, the number shaded would be 20, which is 20%. This proportional reasoning is something adults do on an ongoing basis but do not necessarily know how to explain. In the book *Keeping Things in Proportion: Reasoning with Ratios,* the focus is on proportional reasoning.

Spend time discussing the extension problems about interest, which give students practice in rounding as well as in finding percents.

In the first problem, Rami pays off 10% of what remains.

1. Rami owes $100. If he pays off only 10% of the total amount remaining each month, when will he pay off the whole amount?

Technically he can never pay it off if we don't round. Since we are dealing with coins and rounding he will have it paid off in approximately 80 months because as the amount owed gets smaller so does the 10% off. You can explore the patterns if you enter the amounts in a spreadsheet

In the second problem, students can discuss the impact of paying off a debt on a tighter timeframe (20% a month instead of 10% a month).

Facilitation

Introduce *Activity 2: Display Plans* by sharing a blueprint as a map of space use. Discuss the process of planning—drafting ideas, changing them, and delivering a final plan. This will encourage students to use extra grid paper to plan their work carefully before posting their final results.

When necessary, introduce the concept of market stalls and display space to students by using pictures.

Calculator Methods for Finding Percents

Learning to use the percent key on a calculator can be challenging, particularly if students are using a scientific calculator with "shift" functions. Finding a percent of an amount is presented in these lessons primarily by using mental math strategies. Students connect 10% with its fraction equivalent and so divide quantities by 10 and then multiply to find 20%, 30%, 40%, etc. They also explore using decimal equivalents to find a percent of a number.

Introduce the use of the percent key in the context of a series of lessons on the calculator because finding percents with the "%" key on the scientific calculator is not a trivial matter. The steps for finding percent on a calculator vary with the brand and sophistication of the device. On the TI-30XS MultiView™, 10% is found by entering "80 × 10 2nd % enter." Internet sites that show how to use the TI-30XS MultiView or other calculators could be helpful for students. Checking answers on the calculator is an opportunity for students to articulate the equivalent ways to say and write 10% ($\frac{1}{10}$ and 0.1), but be sure they give equal weight to their own thinking before discounting it in favor of the calculator's answer.

Making the Lesson Easier

Students who struggled with *Activity 2: Display Plans* should work again with a design based on 50 squares in *Activity 3*. You can decrease the number of item types displayed, for example: 60% apples, 20% pears, and 20% grapes. Make sure students can show a percent with an array of 100 squares. Review with students the idea that a percent tells us "how many out of 100" and relate finding a percent to finding a fraction with a denominator of 100, such as $\frac{10}{100}$.

Then ask students show a percent with an array of 50 squares. Help students find percents of 50 by using relationships between familiar percents and familiar fractions: "What fraction is 50%? Yes, it is half, so what is half of 50? What fraction is 10%? It is $\frac{1}{10}$, so what is $\frac{1}{10}$ of 50?"

Also help students to reconsider the idea of percents as part-whole relationships and to use part-whole language when they work with percents, for example, "50 of every 100 balloons are red" or "Two out of 10 is the same as 20 out of 100."

Shading grids and outlining areas is also helpful if students struggle to see that $\frac{20}{100}$ is the same as $\frac{2}{10}$, or $\frac{1}{5}$, or that $\frac{30}{100}$ is the same as $\frac{3}{10}$. If they forget the fraction equivalents, students can always return to visual representations, either mentally or physically with manipulatives.

The following three shaded grids represent three different ways to visualize fractions for 20%.

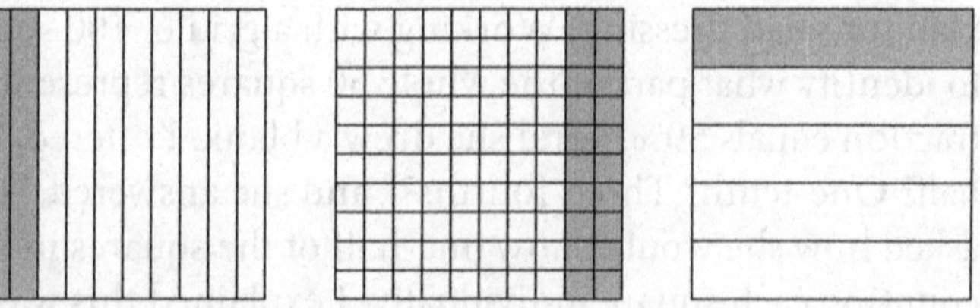

Students learn that every percent includes a number that could be written as the numerator of a fraction that has 100 as the denominator. For example: 20% = $\frac{20}{100}$ or 75% = $\frac{75}{100}$.

Making the Lesson Harder

In *Activity 2: Display Plans*, consider multiples of 5% as well as 10% for display percentages. In *Activity 3: What Is Your Plan?* you can offer grids and a design based on 60 square feet or some other number not so easily marked off in percents.

One student found Activity 3 *a helpful way to think through an entrepreneurial idea of her own.*

Lorna's work was quite interesting this week. She wanted to create her own plan for salad dressing. Working with a grid of 100 squares, she was unable to identify what part of the whole 50 squares represented. I asked her, "What fraction equals 50%?" and she drew a blank. I offered, "One-third? One-half? One-tenth? Three-fourths?" and she answered, "One-half." When asked how she would show one-half of the squares in the grid, she began by counting each square individually. I explained this was one way, but how about looking at the number of rows and coloring in half of them? She was able to do this and subsequently found 30% and 20% easily. However, when I asked her how she knew three rows were herbs, she was unable to answer. She seemed to need more practice articulating the connections between percents and fractions based on 100.

Lorna wound up changing her plan for salad dressing to one for a homemade oil with scents and did a beautiful job. She was pleased that she came up with her own "recipe" and wanted to market her product.

Marilyn Moses, observed by Marilyn Matzko
Brockton Adult Learning Center, Brockton, MA

Shining a light on the trickiest fraction-percent equivalents gave students a chance to reason through the dilemmas.

The teacher wrote the following on the board:

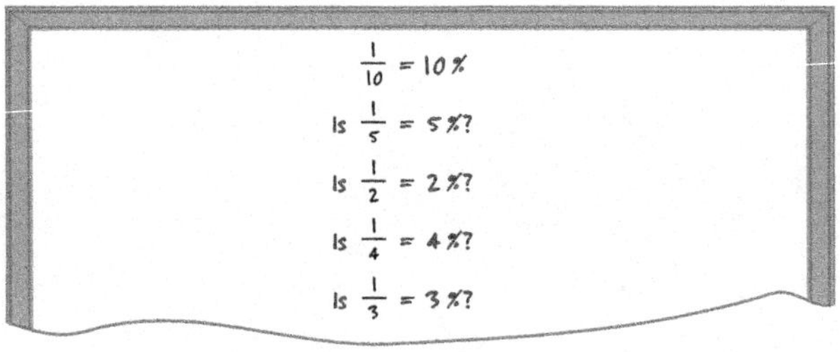

At first, students agreed that one-fifth equals 5%. Even though one person said, "If I have a dollar, then one-fifth is 20¢," the group stuck with five as the percent equivalent.

Carol drew a rectangle split into five parts, and suddenly the class saw that 5% could not be correct because the sum of the pieces was 25%, not 100%, so the percent had to be larger. Then they realized that the comment about 20¢ led to 20%.

continued on next page

continued from previous page

Of course, they quickly disputed ½ = 2% and ¼ = 4%. I added: "Does $\frac{1}{20}$ equal 20%?"

They said "No" right away but did not know what the answer was yet.

Then Ping said, "It's 5% because if $\frac{1}{10}$ is 10% and that is the same as a dime, then $\frac{1}{20}$, which is 5¢, is half, so it has to be 5%." This made sense to everyone.

The next time the class met, I wanted to see what they remembered. "What fraction does 20% represent?" I asked.

Student: "One-fifth."

The others: "Yes! One-fifth."

Teacher: "If 20% equals one-fifth, 40% equals how many fifths?"

Students: "Two-fifths."

Teacher: "And what would three-fifths be?"

Students: "Sixty percent."

Teacher: "Eighty percent would be..?"

Students: "Four-fifths."

Teacher: "And 100%? Five-fifths!"

Carol Kolenik, observed by Myriam Steinback
Harvard Bridge to Learning and Literacy Cambridge, MA

4

FACILITATING LESSON

One Percent of What?

Is 10% always larger than 1%?

Synopsis

In this lesson, students build a foundation for finding any two-digit percent. They increase their percent repertoire, moving from finding 10% and its multiples to finding 1% and its multiples. They also consider the importance of the "whole" when determining percents.

1. Students discuss the meaning of 1% and locate it on a number line before discussing ways to find 1%.
2. Student pairs compare 10% of a smaller number with 1% of a larger number.
3. The class considers patterns related to finding 100%, 10%, and 1% of the same number.
4. The class summarizes what they have learned and expands the equivalents chart in *Reflections.*

Objectives

- Find 1% of two-, three- and four-digit numbers
- Compare 10% of one amount and 1% of another to articulate the effect of the size of the percent and the whole on the size of the answer

Materials/Prep

- Calculators
- Colored markers
- Pennies for the *Opening Discussion* (or colored counters; if pennies are used it helps to have $10 of pennies in rolls)
- Scissors
- Tape

Prepare sufficient copies of *Blackline Master 4: One Percent on the Number Line*, one for each student to use during the *Opening Discussion*, and *Blackline Master 5: Money Strips* for students to fold and mark off as dollars.

Extra copies of *Blackline Master 10: 100-Block Grids* may be helpful.

Opening Discussion

Students report any percents they have encountered since last class, comparing them with multiples of 10%—is the number less than or more than 10%? Is it less than or more than 20%, etc.?

Ask:

We know that 10% is the same amount as $\frac{1}{10}$. What is the fraction for 1%?

Listen for and record responses. Draw a number line marked in tenths on the board. Ask a volunteer to label the tenths and others to add the equivalent percent and decimal terms. The final product should look something like this:

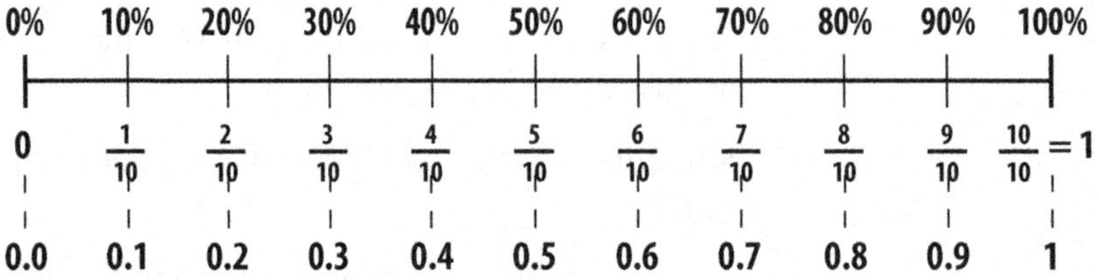

Ask:

Where would you put a line showing 1%? How would you mark it?

Expect students to grapple with the following ideas:

- One percent is less than 10%.
- There are nine marks (10 intervals) between 0% and 10%, one for every one percent: 1%, 2%, 3%, etc.
- One percent is $\frac{1}{10}$ of 10% ; it is $\frac{1}{100}$ of the whole.

Distribute *Blackline Master 4: One Percent on the Number Line.* Invite students to mark the line representing 1% with appropriate terms.

What is the decimal for 1%, or $\frac{1}{100}$?

Point out the relationship between one cent ($0.01) and 1/100 (0.01) if the decimal meaning seems obscure.

Heads Up!

It may be necessary for some students to see and count all the marks to verify that 1% equals 1/100. A meter stick works well for this.

If I wanted to find 1% of $200, for instance, how would I do that?

As a class, "make" 200 dollars using the strips copied from *Blackline Master 5: Money Strips*. Students then show you 10% and 1% of the strips.

How could we split up the $200 to find 1%?

Summarize the processes students used to show 1%. Refrain from formalizing rules until students have more practice with finding 1%.

For students unclear about the idea that 1% of $200 is $2, use the number line to emphasize two ways (using two different scales) of showing the amount: a percent, as if out of 100, and an amount of real money based on $200.

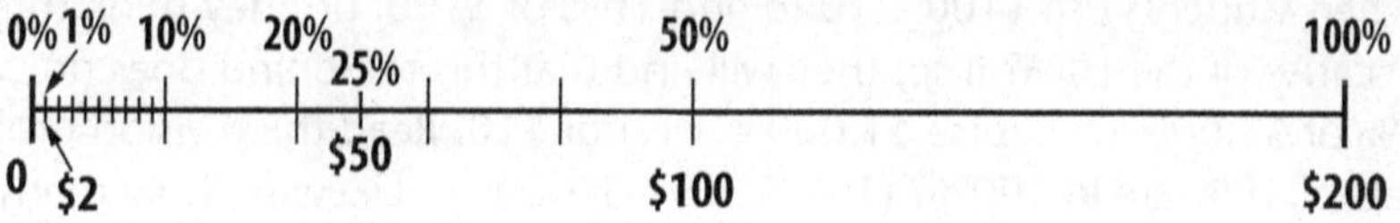

Activity 1: Go Figure!

Students are asked to compare 10% of 100 and 1% of 1,000 as a way to draw attention to the importance of the base amount, the *whole*, from which a percent is taken. Although it may seem that 10% must always be greater than 1%, this activity shows that is not necessarily the case when the whole amounts change.

Review the directions for *Activity 1: Go Figure!* (*Student Book*, p. 62). Pairs choose one of the tools listed—number lines, grids, or diagrams, objects or a table—and a second one to confirm their results (which may include using calculators or pencil-and-paper operations).

Heads Up!

Students often blurt out answers immediately. You might take a quick vote on which amount students think is larger, but then move on. Students will share their findings and their reasoning later.

Allow time for student pairs to work on the problem.

Ask:

What is the whole? Show me on your picture, number line, or grid, etc.

What part of the whole are you looking at? Where do you show that?

Share at least one example of each solution method. Ask:

Where is the 100% in this diagram?

Where is the 10% on this number line?

How do you know this is 1%? What part of the whole does that represent?

How did you decide to label numbers, shade blocks, or arrange objects?

Add:

Which method made the results clearest to you? Why?

If no one used division, ask:

How could you use division to find the percents in each case? ($100÷10 = $10; $1,000 ÷100 = $10)

Heads Up!

When students use $100 ÷ 10 to find 10% of $100, do they think they are dividing by 10 because of the 10%? If so, they will find that this reasoning does not work when finding 1% of $1,000. Of course $1,000 ÷ 1 is not $10. Read the division problem as "How many sets of 10% are in 100%? (10). $100 ÷ 10 is $10. Likewise, how many sets of 1% are in 100%? (100). $1,000 ÷ 100 = $10. This will be explored further in Activity 2 and Lesson 5 In Action shows how the conversation played out in a class.

When strategies are clear, ask:

Ten percent is larger than 1%, right? Then how could 1% of 1,000 be *equal* to 10% of 100?

Expect students to discuss the difference in the size of the amounts, the *whole*, as well as the difference in the percent sizes. If no one makes the point that although 1% is 1/10 of 10%, 1,000 is 10 times *larger* than 100, present the following:

One thousand is 10 times *larger* than 100 (write 100 x 10 = 1,000, or 100 + 100 + 100…), **but 1% is 10 times *smaller* than 10%** (return to the number line demonstration), **so the answers are equal.**

What do you expect will happen if we compare 1% of 100 with 10% of 500?

What do you expect will happen if we compare 1% of 200 with 10% of 1,000?

Allow time for students to solve the problems. If students are still struggling to understand, encourage them to use objects or number lines, such as this number line for 1% of 100 compared to one for 10% of 500:

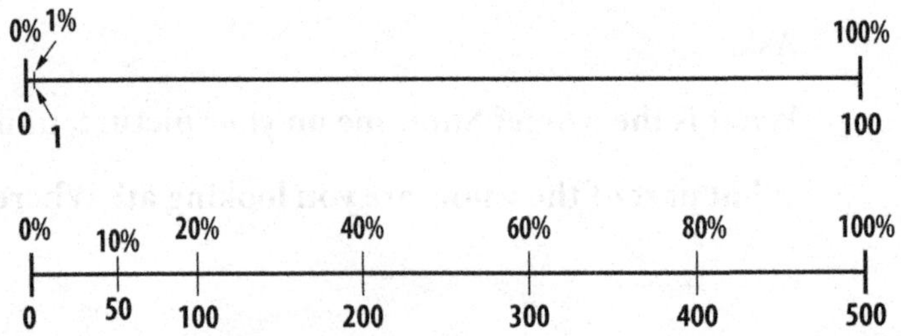

Then ask:

What do you need to consider when you want to predict the answer in a percent problem?

Listen for understanding that the percent amount—the *part* they are finding—is important, but the *whole*—or the original amount—is equally important when comparing percents.

Ask:

Will 1% of 2,000 be greater or less than 1% of 1,000? (Greater. Same percentage so the larger whole will be a greater value). **How do you know?**

One percent of 2,000 equals what percent of 1,000 (2%)**? How do you know?**

For the second question, you may need to prompt students to compare 1% of each whole.

Before moving on, say:

One percent and 10% are benchmark percents. You can use these benchmarks to find the percent of any number.

Activity 2: Patterns with 10% and 1%

Ask:

What is one way to find 10% of a number using division? (Divide by 10.)

What is one way to find 1% of a number using division? (Divide by 10 and then divide by 10 again, or divide by 100.)

Post the methods explained. Say that these division methods will be helpful as the class explores *Activity 2: Patterns with 10% and 1%.*

Refer students to *Patterns with 10% and 1%* (*Student Book*, p. 63). Announce that you will call out a few numbers and then volunteers will suggest some more numbers. Students will determine and record the amounts equal to 100%, 10%, and 1% for each of these numbers.

One or more students may use a calculator. Rotate this role so that everyone has a chance to use the calculator as well as to calculate mentally or on paper. Demonstrate how the calculator can be slower than mental math by asking half the class to use calculators and half to use mental math to find 10% of 200.

Start with three-digit numbers. Using numbers that end in zero, call out some three-digit dollar amounts such as $300, $450, $600, etc. For each number called out, ask the following sequence of questions:

What is 100% of … ? How do you know?

What is 10% of … ? How do you know?

What is 1% of … ? How do you know?

A filled-in table will look similar to this:

Percents	Number $300	Number $450	Number $600
100%	$300.00	$450.00	$600.00
10%	$30.00	$45.00	$60.00
1%	$3.00	$4.50	$6.00

When students grasp this concept, ask volunteers to suggest some more three-digit numbers to complete the table *Student Book*, p. 63.

Begin a second round with four-digit numbers. Call out some four-digit dollar amounts such as $8,000, $3,600, $5,200, etc.

Heads Up!

Start with numbers that are money amounts, as the decimals will make more sense to students. Make sure to write the number consistently with all the zeroes (dollars and cents) to accentuate the visual pattern of shrinking amounts.

After completing three-digit *and* four-digit rounds, ask:

What pattern do you notice as you look down the columns?

What causes the numbers to change as they do?

Students might notice that the numbers going down the column become 10 times smaller in each successive row, or conversely, looking up the columns, the numbers become 10 times larger in each successive row. Some might notice that the numbers in front of the decimal point increase or decrease. Connect these observations with the changes in percents.

Revisit the strategies for finding 1%. Return to the concrete models students showed in *Activity 1* to firmly establish the meaning to students. Then initiate filling in the table with two-digit numbers (these are harder) such as $30, $25, $66, etc.

If students start using the method of "moving the decimal point," ask:

Why does it work to move the decimal point one place (or two places) in this case?

Summary Discussion

Post the term "1%" on the board. Check in with students about the meaning of the symbol:

What can you tell me about 1%?

Listen to students' comments. Ask:

How could you rewrite 1% as a fraction? As a decimal?

Summarize:

"Find half" means split in two or divide by two.

"Find $1/10$, or 10%" means divide by 10.

"Find 1%, or $1/100$," means find a tenth and a tenth of that tenth or divide by 10 twice or divide by 100.

Together post definition information on the class vocabulary list (*Student Book*, p. 204). Then suggest that students expand the equivalent chart and answer the questions outlined in *Reflections*, (*Student Book*, p. 211) as a way to keep track of what they have learned in this lesson.

End with:

Is 1% a little or a lot?

If it is not mentioned, say that although 1% represents a small fraction of an amount, if the whole is large, 1% will be a lot.

Practice

Which Is Greater? p. 64
For practice finding 1% of an amount and comparing percents of different amounts.

Fundraisers, p. 65
For practice finding 1% of various amounts.

Growing Cities and Towns, p. 66
For practice finding 1% of populations and seeing the effect of adding that 1% to the original.

Rounding to the Nearest Whole Number, p. 67
For review of rounding from tenths to whole numbers. Rounding is a useful strategy for estimating in situations that require operations with decimal numbers. Students encounter such situations in later lessons.

Extension

Working Backward, p. 70
Given 1%, students find the whole.

Test Practice

Test Practice, p. 71

Looking Closely

Observe whether students are better able to

Find 1% of two-, three- and four-digit numbers

Do students have a clear understanding of 10%? Students who understand the relationship between 10% and 1% can more easily find 1% of a given amount. Two important ideas may be helpful: First, 10% is the same as $\frac{1}{10}$ and second, partitioning the whole into 10 groups is the same as finding $\frac{1}{10}$. Return to *Lesson 1* practice problems and review 10% concepts before pressing ahead with 1% concepts.

When considering ways to find 1%, do students rely on rules? Rules such as moving the decimal point to the left one place to find 10% of a quantity and moving it to the left two places to find 1% are not invalid. However, if students do not understand the concept and the relationship to division (dividing by 10 or by 100), they can easily fall into traps—moving the decimal in the wrong direction or moving it an incorrect number of places.

Grids are helpful to clarify students' understanding of the fractional equivalent for 1%. The idea that 1% represents "one out of 100" can be shown by shading one out of every 100 blocks on a grid. Start by finding 1% of multiples of 100, such as 300.

Compare 10% of one amount and 1% of another to articulate the effect of the size of the percent and the whole on the size of the answer

Do students understand that 1% equals $\frac{1}{100}$? Understanding that a percent is a fraction helps explain the importance of the whole as well as the part. Make sure students understand the fraction-percent relationship if they struggle with the notion that 10% of one quantity can be larger or *smaller* than 1% of another quantity. To compare parts, percents must always be considered in relation to a whole; one must always ask, "One percent of what?"

By shading grids, marking number lines, and partitioning objects or pictures, students can learn to "see" the fraction-percent relationship. Some students use a mnemonic to help them remember this connection. They see the percent symbol as a division bar and the two circles on either side (%) as representing 100.

Rationale

By starting with benchmark percents—10%, 25%, 50%, and 75%— and combining them with 1%, it is easy to find *any* percent of an amount.

Math Background

This lesson surfaces the concept that percents do not represent absolute amounts, rather they represent part-whole relationships. Newspapers, for example, report percent increases or decreases in various job sectors. In one such report, high-tech jobs showed the greatest percent increase of any job sector, while retail jobs showed greater absolute increases but a smaller percent increase. To understand this difference, one needs to understand the effect of the *whole* on the percent. A percent is meaningless unless considered in context.

When considering the values derived from finding 1% of a quantity, students learn to rely on previous knowledge about finding 10% of a number. The use of 10% and its multiples, as well as 1%, can help in finding any two-digit percent.

Facilitation

In this lesson, the term "the whole" is preferred to the term "the base amount." This preference results from efforts to remain consistent in terminology throughout the *EMPower* books. Students are looking at a whole and finding a part of it, though that part is described as a percent, not as a fraction.

Making the Lesson Easier

Devote more time to comparing 1% of large and 10% of small amounts. Compare 1% of 500 and 10% of 50, for instance. Begin by comparing 1% of 100 and 10% of 50, and work up to 1% of 500 so students can see that as the larger number approaches a magnitude 10 times that of the smaller number, the two solutions approach equality.

Making the Lesson Harder

For *Activity 2*, introduce two- and three-digit numbers that do not end in zero. Also, ask why there is a $1.50 difference between 1% of $500 and 1% of $650.

You might also extend *Activity 2* to include the following:

How would you expect the numbers to change if you listed 0.1% ($\frac{1}{10}$ of 1%) **for the three-digit numbers you have listed in your table?**

Copy numbers from the original three-digit table, then complete the following table with numbers representing 0.1% of each number.

Three-Digit Numbers

Percent	Number ______	Number ______	Number ______	Number ______	Number ______
100%					
10%					
1%					
0.1%					

How would you describe the numbers to a boss or co-worker? Write sentences of your own to explain what 0.1% (1/10 of 1%) **equals for two of the numbers above.**

Examples:

- One-tenth of a percent of 650 dollars equals 65 cents.
- One-tenth of 1% of $500 would be 50 cents.

LESSON 4 IN ACTION

Mental math becomes easier for students over time. This teacher reports how students used familiar percents in combination to find less familiar percent amounts.

I told the students they had to do the problems without using pencils; the math had to be done mentally. We did it as a group, they threw out answers, and there was give and take—"How did you get that answer? Are you sure?"

I gave them a number: 240. I asked, "What's 10%?" Everyone yelled "24!" They knew 5% was 12. Then I asked them what 80% would be. They didn't know, but they eventually figured out what twice 10% would be and subtracted that amount from 100%.

Twenty-five percent was easy for my group; it's half of a half—that is, half of 50%. They've been working on that. For 35% and 40%, they used a calculator. I said, "What would you do if you didn't have a calculator? How could you find 35%?" One student knew it was 25% plus 10%. They all knew how to find 10% because we spent so much time working on $1/10$. They can find 10% with a shift of the decimal point.

Susanne Campagna
Read/Write/Now, Springfield, MA

FACILITATING LESSON **5**

Taxes, Taxes, Taxes

How is the tax determined?

Synopsis

In this lesson, students practice finding multiples of 1% as they consider sales tax charges in various states. Multiples of 10% and 1% are then combined to analyze payroll deductions.

1. The whole class discusses taxes and tax rates encountered in daily life.
2. Students solve a series of problems involving the sales tax charged in different states. Pairs of students check each other's work.
3. The class discusses methods for finding single-digit percent amounts and the connection between percents and parts of wholes.
4. Student pairs solve a problem involving payroll deductions for two individuals.
5. The class discusses methods used to find two-digit percents and then considers using a decimal equivalent and the calculator to solve the same problems.
6. The class summarizes by compiling a list of methods for finding two-digit percent amounts.

Objectives

- Use multiples of 1% to find single-digit percents
- Combine multiples of 10% and 1% to find two-digit percents

Materials/Prep

- Calculators
- Colored pencils or markers
- Play money
- Sticky notes (optional)

Copy *Blackline Master 9: 100-Block Grid* for the class to use during the *Opening Discussion.*

Provide actual sales slips showing item costs and sales tax (from more than one state if possible) to share during *Activity 1* (optional).

Collect payroll stubs (with names blacked out) to share during *Activity 2* (optional).

Opening Discussion

Survey the class by asking:

Who has paid taxes? What kind of taxes? (income tax, sales tax, cigarette tax, gas tax)

What percent tax did you pay?

List some of the figures for your state that go with the examples given.

Be specific in your questioning and have available play money for demonstrations. Ask:

How do you figure sales tax on a $10 item? Who uses another way?

Show us your answer is correct using the 100-Block Grid or the play money.

Have available some copies of the *Blackline Master 10: 100-Block Grids*, and allow time for students to work together in pairs or groups.

Heads Up!

If the percent of sales tax in your state includes a decimal, consider using another tax that does not or using a figure from a neighboring state.

The language "sales tax included" may confuse students. Talk through an example. For instance, the cost of a $10 item that is priced with sales tax included and the final cost of that $10 item if sales tax is added on.

Following are two ways students might demonstrate with the grid:

- Decide how the whole grid represents $10. Divide the grid into 10 columns representing 10 dollars, and mark each "dollar" (column) with the amount of tax (cents per dollar) on it. Then add the 10 figures together.

- Show that each block represents 10 cents. Shade the number of blocks representing the sales tax percent (five blocks for 5%, for example). Then multiply the number of blocks by 10 cents to determine the tax amount.

Ask volunteers to explain their reasoning with their grid.

Did any of you get an answer with the grid that surprised you? How?

Take time to reconcile differences and reach consensus on the total tax paid. Then prompt a review for finding 1% by asking:

What if the sales tax were 11%? How could you use what you know about finding 10% and finding 1% to figure the 11% tax on a $10 item?

Record strategies.

Introduce *Activity 1 Different States, Different Charges* (*Student Book*, p. 74) by saying:

Today you will look at several situations involving taxes. The first situation involves figuring out the sales tax charged in different states for three items.

You will see that you can use what you know about finding 10% and 1% of a number to find other percents.

Activity 1: Different States, Different Charges

Refer students to *Activity 1: Different States, Different Charges* (*Student Book*, p. 74). The additional practice in determining multiples of 1% will be valuable. Review directions.

Together determine what the rounded item amounts will be, and ask each student to record that information on the "Sales Tax Table." Assign pairs to check each other's work. Students complete the table and answer the questions on their own.

Pay attention to the following:

- Do students start with 1% multiplied or added? Or 5% because they can find 10% easily and then divide by two?
- Do they build percents by finding 4% (either by determining 1% and multiplying it by four or by finding 2% and then doubling it)? Do they then add 1% to find 5% or 2% to find 6%?

Some students will always begin by finding 1% and then adding or multiplying to find other single digit percents. Some will move the decimal point, while others will divide by 10 and then divide by 10 again to find 1%; some may use a calculator and divide by 100.

Talking about the varied strategies will allow you to connect these different approaches. You want students to see that the decimal point movement results from dividing and that dividing by 10 and then by 10 again results in the same answer as dividing by 100.

When everyone has completed at least the third column for Problem 1, call the class together. Say:

I saw many different approaches to finding the sales tax amounts. What was one strategy that you used to find 5%? Who used another way?

Connect the strategies whenever possible, discussing in detail why both ways work. For instance, point out that to start with dividing by 10 (finding 10%) means you need to divide further because 10% is *more* than 5%, while to start with dividing by 100 (finding 1%) means you need to multiply because 1% is *less* than 5%. However, in both cases the first step requires division.

Keep sharing. Then ask:

Which of these strategies is easiest to do in your head?

Which of these would be easiest to use if the whole were $9,750, not $100 or $40?

Students will likely have varying opinions about "easiest" methods. It is not important to reach consensus here; rather you want students to realize that different strategies may be easier in certain cases, although they yield the same results.

When you ask students how they determined the final cost of a sound system (sales tax included) in Florida, point out the need to add two amounts—tax and cost—in order to find the solution.

Problem solving remains the focus when you discuss how much more a sound system costs (sales tax included) in Michigan than in Alabama. Highlight the need to find total final costs (by adding the amount of the tax to the price of the item) in both states before subtracting to find the difference. On Problem 4, you can also move students toward a generalized rule about percents: The higher the percent charged, the higher the amount paid. Ask:

Is it true that the higher the tax percent, the more you pay? Why?

If students do not mention that the higher the percent charged, the larger the part of the whole being considered, ask about the part/whole relationship.

Close the discussion by asking:

How could you use what you know about finding 1% of a whole to find *any* percent?

Expect students to arrive at the conclusion that they can divide the whole amount by 100 (or by 10 and then 10 again) and multiply by the number of the percent. Check the rule by practicing on a few amounts such as 250, 1,000, and 5,475, using calculators as well as mental math.

Activity 2: Take-Home Pay

Introduce the activity by telling students:

Sales taxes are only one type of tax. Almost all taxes, though, are described in terms of percents paid, as you will see in the next activity which looks at payroll deductions and take-home pay for two people.

Refer students to *Activity 2: Take-Home Pay* (*Student Book*, p. 76). Read the first problem and take a quick poll on whether students agree that Mara would be taking home four times as much pay as her brother.

Students work individually or in pairs. Notice how students determine percents that involve multiples of 10% and 1%.

When students have completed the table, ask:

What did you discover about the take-home pay for these two people?

Ask students to justify their discoveries by explaining why it turned out that the more highly-paid person took home about three times as much money even though she earned four times as much as her brother.

Then turn the discussion to calculation of percents:

How did you find 28% of $1,600?

How did you find 15% of $400?

Did anyone use a calculator? How?

Discuss the methods used for each problem. Suggest alternative approaches if everyone solved the problems the same way.

How might knowing one-fourth, or 25%, of $1,600 ($400) **have helped you solve the problem?**

Focus as well on the whole and the part:

What was the whole in Mara's case? ($1600) **In her brother's case?** ($400)

What part was paid for federal income tax?

If students report the part as 28%, ask:

What fraction is 28%? How do you know?

How can you find 28/100 of $1,600 using the calculator?

Once students have realized they can find the answer by dividing $1,600 by 100 and then multiplying by 28, practice with a few more dollar amounts. Double-check by using non-calculator methods for finding 28%.

If students are using multiplication, reinforce this method by saying:

How could you use the decimal 28 hundredths (0.28) **to find 28% of $1,600 on the calculator?**

When students understand that they should multiply, revisit some earlier problems to employ this method. You might also practice calculating with some different percents such as 37% or 82%. Always prove that results are accurate by checking with the method of finding 10% and 1% and their multiples and then either adding or subtracting.

Establish as a class that to find 28% on the calculator you can do one of the following:

- Divide by 100 and multiply by 28.
- Divide by 10 and then 10 again and multiply by 28.
- Multiply by 0.28.

Summary Discussion

Say to the class:

Sometimes you only need a rough idea of what percent an amount represents; however, other times it is not enough to say, "This is about 10%." Sometimes it is important to know an exact amount, for example, in the case of a 15% tip or a 35% deduction.

What are all the ways you can find 15% of a number such as $400?

Record methods, adding these if no one mentions them:

- Find 10% (divide by 10), then find half of that, and add the two together: $40 + $20 = $60.
- Find 10%, take half of that (5%), and then triple that amount.
- Find 1% (divide by 100), and multiply by 15: $4(15) = $60.
- Find 10% (divide by 10), and then divide by 10 again to find 1%. Add the 10% number and multiply the 1% number by five: $40 + $4(5) = $40 + $20 = $60.
- Change the percent to a decimal (15% = 0.15), and multiply $400 by 0.15 on a calculator.

Reiterate that although using a calculator often appears easy, knowing how to find a percent by using multiples of 10% and 1% is useful when you do not have a calculator handy.

Then ask:

In what ways have you thought about parts and wholes in this lesson?

Suggest students take a few minutes to write in *Reflections* (*Student Book*, pp. 211-212) about two ways to find a particular percent of a given amount and finish the *Equivalents Chart* (*Student Book*, p. 210), if they haven't already. Direct students to *Vocabulary* (*Student Book*, p. 204) and ask them for any terms they would like to add to the list.

Practice

Personal Payroll Deductions, p. 77
For practice finding tax percents deducted from a weekly income.

Which Is a Better Deal?, p. 79
For practice finding percents and fractions of amounts charged for sale items.

Increase, Decrease, p. 81
For practice determining percents and adding to or subtracting from an original amount to calculate the increased or decreased amount.

Visuals of Percents, p. 82
For practice representing the whole when given a part or percent of it.

Mental Math Comparisons, p. 84
For practice calculating percent amounts for multiples of 1%.

Extension

Markdowns, p. 85
Tells the story of how an item price changed due to percent reductions.

Test Practice

Test Practice, p. 86

Looking Closely

Observe whether students are able to

Use multiples of 1% to find single-digit percents

Do students recall how to find 1% of a number? If students struggle to find 1%, ask how they find 10% and how that might help them find 1%. Review the concept of multiples by asking, "If you know that $10 is 1/10, or 10%, of $100, how would you find 30%?"

Students can separate whole amounts into percent parts using play money, or they can use a grid to better visualize the problem. For instance, if a whole grid represents $40, what is the value of each square on the grid? (40 cents) A 100-block grid also represents 100% of an amount, so each block represents what part/percent? (1%) How many must be counted to find 4%? (Four) How much money is that? ($1.60).

Do students know they need to add sales tax and item cost together when comparing costs among states? The introduction of store sales slips can be used to demonstrate this idea.

Combine multiples of 10% and 1% to find two-digit percents

Do students know how to **decompose** numbers? Help students see that they can break 15% into two parts—a 10% part and a 5% part. You can use a 100-block grid to show this by shading 10% in one color and 5% in another and asking what the total percent shaded is. Suggest students take apart 28% as well. Some ways to break apart the number 28% follow:

- 10% + 10% + 8%
- 25% + 3%
- 20% + 8%

If students break apart numbers in ways that do not simplify the problem ("28% = 15% + 13%," for example), suggest that they concentrate on breaking apart numbers into 10's and 1's. Provide them with some practice so that breaking apart numbers this way becomes automatic.

Those who easily break apart numbers and who know how to find 10%, 1%, and their multiples can deepen their number sense by considering:

- How can a benchmark fraction/percent help in this case?
- Is there any way to use doubling or halving?
- If I round up (to 30%, for instance), how would I find 28%? (28% is 2% less than 30%, and I can find 30% and 2%.)

Rationale

Knowing how to combine 10%, 1%, and their multiples helps students determine less common percents, such as 28%. In situations in which an approximation is not good enough, knowing how to find a percent by using the benchmark percents 10% and 1% in combination with 25%, 50%, and 75% is useful.

Math Background

Finding a percent of an amount is only the first step toward determining the full cost of an item or the net take-home pay for an individual. The problems in this lesson are two-step problems; they require a first step, finding the amount of the percent(s), and a second step, adding or subtracting that amount from the whole to find the answer. Keeping track of steps in an organized way helps students solve the problems.

Deciding what operation to use when finding percents will vary. For example, 28% of 50 can be solved by multiplying: 0.28 x 50 = 14; or it can be solved by finding 30% of 50 (15) and subtracting 2% of 50 (1), so 15 – 1 = 14; or by finding 25% (¼) of 50 (12.5) and adding 3% of 50 (1.5), so 12.5 + 1.5 = 14. The preferred method depends on the individual. For those who know how to find multiples of 10% and 1%, for example, finding 30% and 2% and subtracting might be easiest; for those who like using the calculator, using the decimal for 28% and multiplying might work best.

Context

Tax charges commonly surface in daily life. This lesson provides a springboard to discuss taxation, earnings, and state tax rate differences, as well as processes for determining percents. Students can locate information to compare state and local taxes of any type by using an Internet search engine and typing in the type of tax (income, sales, excise, estate, etc.) and the level of government (state or federal) or the locations (cities and towns) they wish to examine.

Facilitation

Whether or not students accurately calculate a percent, you want them to consider how sensible their calculations are. Ask, "Is it supposed to be less than 10%? What would 10% be?" Or, "Is that close to 25%? What would one-quarter be?"

The decimal-equivalent discussion at the close of *Activity 2: Take-Home Pay* opens the door to further exploration. You might choose to explore decimal patterns to construct a general rule for changing percents to decimals. Encourage students to reexamine the percent, fraction, and decimal table on p. 210 in their *Reflections* section. They can add the decimal equivalents for 15% and 28% and then look at all decimals and percents to notice the decimal-point pattern (moving the percent's decimal point two places to the left). Once they see this pattern, you might follow up by including some single-digit percents and their decimal equivalents.

Making the Lesson Harder

Following Problem 4 in *Activity 1: Different States, Different Charges*, you might ask questions that provoke further generalizations, such as:

If the Michigan sales tax were 9%, would you expect the difference between the cost of an item there and of the same item in Alabama to be greater or smaller? Why?

If the Michigan sales tax were 9% and the Alabama sales tax were 6%, how would the difference affect a one-time purchase like a refrigerator? What about something you buy weekly or monthly like paper towels?

Base calculations on the $100 sound system price to make differences easily visible and comparable.

You might also ask students to consider the difference in tax paid on two items—sneakers and sound systems—in states with 4%, 5%, 6%, and 7% tax rates. They look for a pattern, explain it, and, if possible, write a rule for finding sales tax for any item. Chart the differences in taxes paid and post the numbers on the board:

Sales Tax Charged	Difference between Tax Paid on Sneakers and Tax Paid on Sound System
4%	$2.40
5%	$3.00
6%	$3.60
7%	$4.20

Then pose the question:

As we increase the sales tax by 1%, the difference between tax paid on the sneakers and tax paid on the sound system increases by the same amount each time. Why is the difference always the same?

The class will calculate the pattern of increase (60 cents for each percent increase), but may need to be prompted to consider what 60 cents represents (1% of the $60 difference in cost between a stereo and sneakers).

How might you write a rule that would let you predict the difference in tax paid on sneakers and sound systems between any two states?

LESSON 5 IN ACTION

Figuring out percents by using patterns is a powerful method; however, students often need multiple opportunities to do this and to generalize from what they have done to make this method their own.

Teacher: "To find 1% of a number you can . . ."

There was total confusion, although students were well able to find 1% of 500, as well as 10%, 25%, 50%, 200%, and 300%. Students were confused about what they needed to do—multiply or divide.

The teacher wrote on the board: "50% of 500 = 250"

Teacher: "How did we get that?"

Students: "Took half."

Teacher: "Okay, that is the same as dividing by what?"

The students called out, "Two." Then they continued to work on the operation required by each percent. Students listed the equivalents—50% with 1/2, 25% with ¼, 10% with ⅒, and 1% with 1/100—and saw the relationship between each of those percents and division by 2, 4, 10, 1, and 100, respectively.

Having established the list, Carol asked, "How can we figure out what is 1% of 3,200?"

Students worked with percent patterns, starting with 100% = 3,200.

Student: "To get 1%, I knock off two zeroes. That gets me from 100% to 1%, so I knock off two zeroes and that gets me from 3,200 to 32."

Teacher: "What about 1% of 5,000?"

Kiki: "One percent of 5,000 is 4,950."

The teacher took a moment to ask Kiki how she got 4,950. In fact, she found 1% and subtracted. Because she confused 1% of 5,000 with 1% off, she came up with the wrong answer. The group did a few more examples, clarifying each time whether the 1% was *of* the whole or *off* the whole.

Carol Kolenik, observed by Myriam Steinback
Harvard Bridge to Learning and Literacy, Cambridge, MA

Lesson Discussion:

Teacher Writes:

50% of 500 = 250

S: ÷ 2

T: $50\% = \frac{1}{2}$

$25\% = \frac{1}{4}$

$10\% = \frac{1}{10}$

$1\% = \frac{1}{100}$

FACILITATING LESSON **6**

Decimal Hundredths

How do these compare?

Synopsis

Students move from focusing on tenths to focusing on hundredths. They draw upon their prior knowledge of benchmark fractions, decimals, and percents to figure out whether given numbers are equivalent. They make choices about which zeroes are optional and which are not, and they consider the role and meaning of decimal points in numbers. The class builds on benchmark fractions by extending the fraction strips to include decimal equivalents to the hundredths.

1. Students begin the hundredths work by building on the sets of fraction strips they used to explore tenths.
2. Students focus on place value to the hundredths and explore the role of zeroes.
3. The class examines and sorts a group of numbers (fractions, decimals, and percents) and determines which are equivalent.
4. Student pairs examine data on runners' speeds and consider the effect of rounding on the outcome of the data.

Objectives

- Create visuals to show decimal place value in the tenths and the hundredths
- Connect benchmark fractions to decimal equivalents to the hundredths
- Make informed decisions about keeping or dropping zeroes and decimal points

Materials/Prep

- Calculators
- Meter stick
- *Blackline Master 6: Ten Cards*

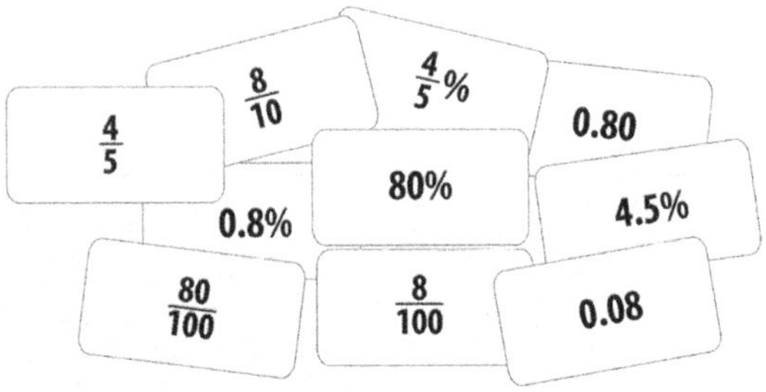

For *Activity 2,* post a place value chart to which examples can be added (See *Student Book, Activity 2,* p. 90 for an example).

For *Activity 3*

- Make one set of cards for each pair of students
- Enlarge one set of cards to use for the board
- Draw a one meter-long number line marked at 10 cm intervals (don't discard it; save it for *Lesson 7*)

Opening Discussion

Start with a reminder of what students explored in *Lesson 2*—the difference between 25 and 52 and 97.9 and 99.7.

Heads Up!

As you talk about numbers, remember to express them formally, e.g., 97 and nine-tenths, as opposed to 97.9. Keep the meaning of the decimal and its relationship to a fraction prominent.

In *Lesson 2,* you answered Hard Questions—a father responding to his children's questions about 97.9 degrees compared to 99.7 degrees and 25 compared to 52.

In this lesson, we are going to examine numbers that have one or two decimal places and explore what those mean. Turn to a partner and give each other an example of a number you know with two decimal places.

As students are sharing, listen for examples you want to make sure get shared with the whole group.

Example: I heard you say that you paid $6.89 for a bagel sandwich. What decimal places were you talking about?

As students explain that the decimal was the "point 89," probe more.

So, I'd like someone to explain to us what exactly that .89 is.

The goal is to have them recognize that "point 89" is the 89 of 100 cents that make the dollar; it is a decimal to the hundredths place.

Today we will explore decimals and make sense of what they mean. We'll discuss how to distinguish between various decimals, and we'll decide how to tell whether we are talking about tenths or hundredths.

Activity 1: Building on Fraction Strips

Refer students to *Activity 1: Building on Fraction Strips* (*Student Book,* p. 88). Explain that this activity is similar to what they did in *Lesson 1,* but this one asks them to make connections to the hundredths.

Ask students to work with a partner to complete the handout. Then bring everyone back together to discuss any issues and discoveries that they made about decimal equivalents.

Be sure that students understand that tenths can be written as hundredths (in Part 3). Students may mention money as an example of when you have to use two decimal places. In *Activity 4,* they will see another example: a race in which each runner is clocked to the nearest hundredth of a second. If we use 4.3 instead of 4.30, some may think that the 0.3 has been rounded. When we use 4.30, then we make it clear that the hundredths have not been rounded to the nearest tenth.

Activity 2: Extending the Places to the Hundredths

Refer students to *Activity 2: Extending the Places to Hundredths* (*Student Book,* p. 90). Encourage them to work with a partner to decide who is right.

Then bring everyone back and review each of the responses. Be sure that students agree that each of the responses could be considered correct, depending on the circumstance. (It would be inappropriate to say 0.70 is seven-tenths if you have been asked to describe an amount in terms of hundredths.)

Summarize, making sure that the idea of differentiating between tenths and hundredths has been understood. Make sure students know that the number to the right of the decimal is tenths and that if there are two numbers, we read them as hundredths.

Confirm understanding by drawing a place-value chart on the board and asking students to help you complete it. Remind them:

In .90 and .09, the first digit tells how many tenths and the second, how many hundredths. The 9's and 0's have different values because of where they are in relation to the decimal point.

Then ask:

What is different about the names to the right of the decimal as compared to the names to the left of the decimal? (Students might observe that there is no units or ones place to the right of the decimal; instead, the first place is tenths. They may also note that we need "th" on the end of the word.)

What is the consistent pattern no matter what the names are? (Each place has a value ten times greater than the place to its right, or one tenth of the value of the place to its left.)

How would you write 3245.82 in the place-value chart?

Practice with as many such questions as students need to feel comfortable. Allow time for students to complete the questions in *Activity 2.*

Activity 3: Finding the Match

Get students thinking about fraction, decimal, and percent equivalence and nonequivalence by saying:

You have to watch carefully when you are working with fractions, decimals, and percents. Sometimes things look the same but are not. Sometimes they look different but are really equivalent.

Form student pairs. Refer students to *Activity 3: Finding the Match* (*Student Book,* p. 91). Distribute a set of cards from *Blackline Master 6: Ten Cards* to each pair. Also display these same ten cards on the board as shown:

	$\frac{4}{5}$%	$\frac{4}{5}$	0.08
	80%	$\frac{8}{10}$	$\frac{8}{100}$
$\frac{80}{100}$	4.5%	0.80	0.8%

Examine these numbers and choose a way to sort them. Then see whether you can think of a different way to sort them.

Ask students to share first with a partner. Then share together as a group.

There are several ways to sort the numbers. Some students might have sorted by format, putting together all fractions, all decimals, and all percents. They might share this method by saying:

- "0.8%, 80%, 4.5% and 45% are all percents. Their values, though, are not the same."
- "0.08 and 0.80 are both decimals. "
- "$\frac{4}{5}$, $\frac{8}{10}$, $\frac{8}{100}$, and $\frac{80}{100}$ are all fractions."

Others might have sorted by value, matching equivalents they notice.

For example,

- "80% = $\frac{80}{100}$ = 0.80. They all mean 80 hundredths."
- "$\frac{8}{10}$ = $\frac{4}{5}$, which is the same as 80%."
- "All the rest are each a different value, so I grouped them by 'equal to 80%' and 'not equal to 80%'."

Focus on one of the groups that have have sorted by value. If no one has sorted by value, ask students to do that now.

Ask students to look at the one-meter-long number line you've posted on the board or wall. Mark the line at 10 cm intervals.

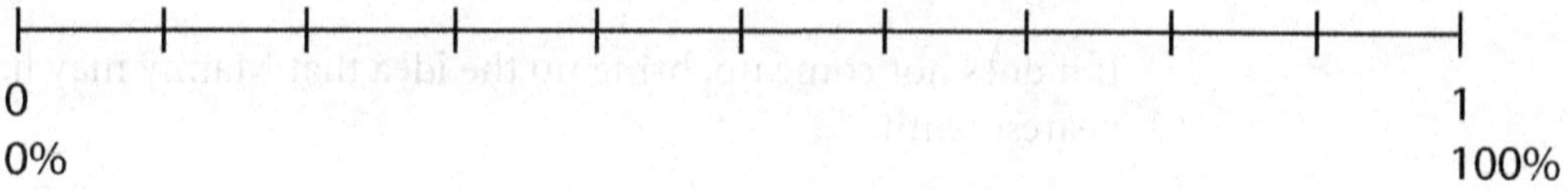

Here is a number line that begins at 0 and ends at 1. We can think of it as starting at 0% and ending at 100%. Where on the number line does each card belong?

What is the value of each of the small marks along the way? (Label them $\frac{1}{10}$, $\frac{2}{10}$, etc.)

How could you say those fractions as percents? (Label the number line 10%, 20%, etc.)

Ask for volunteers to come up, one at a time, to place a card on the number line. Each time, ask the rest of the class whether they agree or whether someone can make a convincing case for moving the card. This should spark some good discussion about equal values.

The final result is to see that five of these numbers are equivalent and have the largest value (80% = $\frac{8}{10}$ = $\frac{80}{100}$ = $\frac{4}{5}$ = 0.80). Numbers are equivalent when they occupy the same places on the number line. The rest, starting from the smallest values, are: $\frac{4}{5}$% = 0.8% (which are both less than 1%). Then, 4.5%. Then, $\frac{8}{100}$ = 0.08. Save this number line with cards for the *Opening Discussion* in *Lesson 7*.

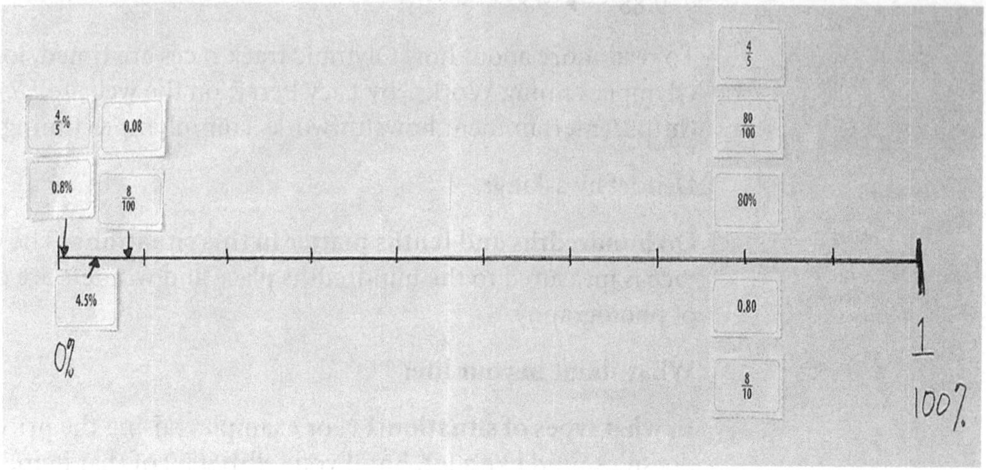

Activity 4: The Runners' Dilemma

Begin the discussion by asking students about races they have participated in or heard of. Then explain that this activity reinforces the need to be able to talk about and use very small amounts (such as tiny portions of time). Ask:

How long is a second? (1 sec. is the time it takes to say "one Mississippi")

How long is a tenth of a second? (0.1 sec. is just a blink or two)

How long is a hundredth of a second? (0.01 sec is about $\frac{1}{40}$ of a blink)

Let students work in pairs or triads to discuss the dilemma. When everyone has completed the tasks on the *Activity 4: Runners' Dilemma* (*Student Book,* p. 92), bring everyone back together and discuss their results.

If it does not come up, bring up the idea that Manny may have rounded to the nearest tenth.

9.63 → 9.6

9.75 → 9.8

9.79 → 9.8

9.81 → 9.8

9.88 → 9.9

Luis focuses on the hundredths of a second, just as the times are written.

For Don, the justification for calling a tie might be that the race is thought of in tenths of a second, rather than hundredths. So if your clock measured only tenths of a second, 0.70 through 0.79 seconds would not appear. What you would see for those times is just 0.7 seconds. In this case, the runners' times would look like this:

9.63 → 9.6

9.75 → 9.7

9.79 → 9.7

9.81 → 9.8

9.88 → 9.8

To read more about how Olympic track races are timed, look for the article "How Olympic Timing Works" by Lacy Perry, on the website *How Stuff Works.* (http://entertainment.howstuffworks.com/olympic-timing.htm)

Debrief by asking:

Do hundredths and tenths matter in this situation? (The Olympic 100-meter race is measured to the hundredths place and winners are determined with the aid of photography.)

What about in your life?

In what types of situations? (For example, raising the price of an item by 1¢ doesn't sound like a lot, but if you sell 10,000 of that item, it matters. That's $100.

Direct students to the *Reflections* (*Student Book,* p. 212) and to *Vocabulary* (*Student Book,* p. 204) to summarize their thoughts before they complete the practice pages.

Summary Discussion

Ask students to share what they learned from the activities in this lesson:

What does equivalence mean? (Two numbers have the same value, but not necessarily the same form.)

How can you tell that two numbers are equivalent? (Probe for various methods.)

How would you compare a digit in the hundredths place with the same digit in the tenths place?

Zeroes are sometimes very important and other times not necessary. Give an example of when the zero is not important. Give an example of when the zero is necessary.

Practice

Designer Accessories, p. 93
For practice allocating space using percents for one-eighth and three-eighths.

Round Up or Down?, p. 95
For practice finding 12% and 13% and comparing them to 12½%.

Better Deal?, p. 96
For practice finding 30%, 33.33%, and one-third using division by three.

How Much, How Far? p. 99
For practice using parts and wholes to figure thirds and percent equivalents.

Extension

Budgeting, p. 100
Students rely on benchmark fractions and percent equivalents learned thus far.

Test Practice

Test Practice, p. 102

Looking Closely

Observe whether students are able to

Create visuals to show decimal place value in the tenths and the hundredths

Visuals for explaining hundredths include 10×10 grids and number lines marked in increments of .05 or $\frac{5}{100}$. A real-world example might help students. They can picture a meter stick as one whole with each centimeter counting as one-hundredth of the stick, or a football field as one unit, with each yard being one-hundredth of the field. In urban areas, students might be able to imagine each foot of the 100 feet between streetlights as one-hundredth of the distance. Ask if they buy anything (child-care supplies, tiles or other construction materials, for example), in sets of 100. In whatever way students decide to make their sketches, look for equally divided parts and an understanding of the relationship of tenths and hundredths, parts and wholes.

Connect benchmark fractions to decimal equivalents to the hundredths

Periodically (or whenever possible) confirm that learners can picture decimal quantities written or sketched as a fraction, as parts of a whole. Using the number line is one way to anchor students' understanding of parts and wholes while they explore the sub-division of tenths into hundredths. Listen for students' understanding and appreciation that decimals are another way to write fractions. Being able to move back and forth between decimals and fractions means students will more easily be able to compare quantities, check themselves using mental math as well as calculators, and find percent equivalents. Each of these approaches can have benefits depending on the context, so flexibility is key.

Make informed decisions about keeping or dropping zeroes and decimal points

When students have a good grasp on decimals, they understand the purpose of zeroes. In some numbers, zeroes are essential to show there are no tenths, for instance, in $10.08. In other numbers they may be optional, so that we might record them for convenience—for instance, when lining up two quantities to compare them, as in 0.09 versus 0.90. If students continue to struggle with this, make up a list of numbers and ask whether zeroes can be added or inserted in various places and why or why not. Reference *Lesson 2, Math Inspection: Entering Numbers into Machines (Student Book,* p. 24).

Decimal points and the numbers after them, including any zeroes, give us a sense of accuracy. The more places indicated on the right of the decimal point, the more accurately measured the quantity is. Students should feel that they have the right to use their discretion, weighing the meaning of their actions if they choose to round off. Listen for contexts where accurate measuring might matter to students and their well-being and to times when they are able to make thoughtful decisions about rounding. Such contexts are often tricky because an amount to the thousandths place is often hard to perceive. The impact of a thousandth of a penny accumulates at larger scales than most of us can easily imagine or appreciate. Utility companies benefit from this. Most consumers will ignore an increase in fractions of cents, as too small to protest. But over time and thousands of customers, these tiny amounts add up.

Rationale

Working with decimals to the hundredths place presents an opportunity to go more deeply into an area where students have some familiarity. Nearly everyone handles money and counts change, though the relationship between units can often go unexamined. In this lesson, one point is to make the meaning of the decimal place clear. That is, 0.02 is to be read and understood as "two hundredths" and to be distinguished from 0.20 which is "twenty hundredths." Another important point is to understand the relationships among decimal places so that students can use that as they reason about operations and compare numbers. This lesson lays the groundwork for estimation and operations with decimals because students begin to understand the purpose of the decimal point.

Math Background

Decimal numbers often create challenges because of the decimal and the names of the decimal places. One of the critical aspects of working with decimals is to see that decimals are numbers that include a bit more detail than whole numbers. We use them every day to deal with amounts of money, and they also come into play in lots of measurements, including those of time, distance, length, width, height, and weight. If students are able to see decimals in those contexts, they will make sense of them.

Issues that arise—"what happens when the zero is to the left or to the right of the decimal?" and "why are 0.7 and 0.70 the same?"—are key. You will have opportunity to explore them in this lesson.

Facilitation

When drawing analogies to money, make sure to also name the decimal in its formal way (89¢ and 89-hundredths) to keep the meaning of the amounts, the place value, and decimal notation all connected. Meaning is lost when we say "point eight nine."

Having worked with zeroes in *Lesson 2* with jackpots in the millions, etc., students may feel comfortable and accustomed to long strings of zeroes. In any case, keep checking that students can explain why zero is in the middle of a number and how that affects the quantity as well as how you read the amount.

Making the Lesson Easier

Use examples with money. Have in mind the real cost of items, measurements students are familiar with (e.g., temperature of 98.5; 2.75 lbs. of chicken; 3.5 yards of fabric, etc.) as a way to connect numbers to familiar amounts. Have them write as well as say the numbers aloud, "three and five tenths"; "ninety-eight and five tenths."

Making the Lesson Harder

For extra challenge, students may sort the cards in *Activity 2* in a third way, looking for other patterns or similarities.

Assign *Extension: Budgeting* (*Student Book*, p. 100) without limiting the fraction types and ask students to allocate some percent or fraction for savings each month.

The students connected benchmark fractions and percents to decimal equivalents to the hundredths. The teacher fostered her students' mathematical thinking by asking questions. The teacher started by explaining that fractions and decimals and percents look the same but are NOT equal. Sometimes they are equal but don't look the same.

In pairs students sorted cards into groups.

A student said he was confused about $\frac{4}{5}\%$. The teacher asked questions, such as: "Can $\frac{4}{5}$ and $\frac{4}{5}\%$ be the same?" "What about $\frac{5}{5}\%$? What would that equal?" "How do these amounts relate to dollars and cents?"

Then the teacher wondered: "What about 4.5%? Is this the same as 80%?"

After the card sort, the teacher posted a number line:

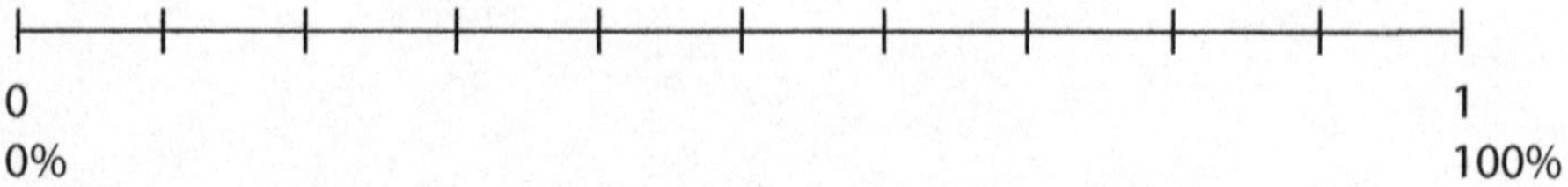

Together the students labeled the incremental lines with percents. Volunteers came up to label each line with a fraction. For each label, the teacher asked the class to answer how they knew or how they could explain their decision. Then students went up and put their cards on the board. They were very involved. All were looking at their individual cards and they were talking to each other about the problems. They usually came to consensus on the placement.

The teacher then asked students to summarize: "What can you say about the cards that have the same value? Are they interchangeable? For example, if I multiply with $\frac{4}{5}$ and 80% will I come out with the same answer?"

The teacher noted that she will ask this question again in upcoming classes until all students are certain of the answer.

Amity Gottschalk, observed by Betsy Hill
Bronx Adult Learning Center, New York City, NY

7

FACILITATING LESSON

Smaller and Smaller

How do I express 3 ½ cents with a decimal point and a dollar sign?

Synopsis

Students move to a focus on thousandths, and beyond. They explore visual models and various contexts and draw upon their prior knowledge of benchmark fractions, decimals, and percents to figure out whether given numbers are equivalent. They also practice using expanded notation (writing a number to show the value of each digit).

1. Students begin to consider the placement of 0.008 on the number line and extend the place-value chart beyond the hundredths place.
2. Given three random digits and a decimal point, students create decimal fractions with values closest to common benchmarks.
3. Students interpret and extend information on a set of cards that represent different situations in which thousandths are used.
4. Students express a decimal fraction's value with expanded notation.

Objectives

- Compare thousandths to hundredths, tenths, and ones
- Write decimals in expanded notation
- Round decimals in the thousandths to the nearest 1, 0.1, and 0.01

Materials/Prep

- Calculators
- *Blackline Master 7: Thousandths Cards* one set of cards in an envelope , one per student
- *Blackline Master 8: 0-9 Cards* for *Activity 1*, or preferably 10-sided dice (one per student or per pair of students)
- Copy of an electric bill or students' own electric bill for *Extension: Rate Hikes—Ouch!* (optional)

Opening Discussion

Open the discussion with the number line display that resulted from *Lesson 6, Activity 3* (*Student Book*, p. 91). Have the display handy. On the board, write 0.008.

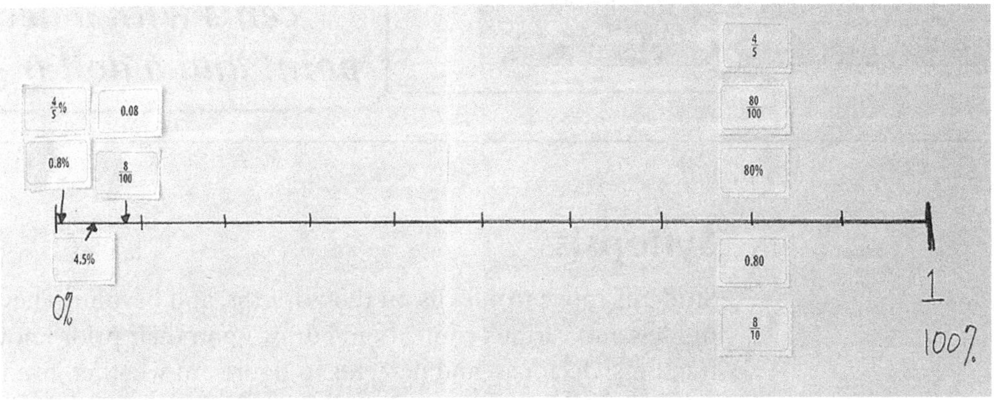

Where does this number belong on the number line? Why?

Have students share opinions. Record all on the board. Then continue to probe, referring to the place-value chart, placing 0.008 to show **8/1000**. Ask:

What relationships do you notice in the place-value chart? How does the value of 8 change?

8 0.8 0.08 0.008

We have primarily been working with decimal values that we understand as dollars-and-cents. Let's think about decimals that have some digits with very small values. By the way, the digits alone don't determine the values, the number of digits influences the value.

Together read the introduction to *Lesson 7* (*Student Book*, p. 103) and add the word "**thousandths**" to the *Vocabulary* (*Student Book*, p. 205).

Activity 1: Target 1, 1/2, and 1/4

Refer students to *Activity 1: Target 1, 1/2, and 1/4* (*Student Book*, p. 104).

The two rounds of this activity involve decimal subtraction and target the numbers 1, ½, and ¼, using either a ten-sided die or the set of number cards from *Blackline Master 8: 0-9 Cards.*

When students finish, ask:

Which number form did you find easier to work with as you found your target? Why?

What strategies did you use?

Activity 2: Smaller and Smaller

Provide each student with an envelope of cards from *Blackline Master 7: Thousandths Cards.* They will keep these for use in several activities.

Ask students to do a quick examination of the cards to see if there are any situations that they have seen in their lives. After having them share general thoughts on the cards, ask them to focus on *Activity 2: Smaller and Smaller* (*Student Book*, p. 106). Ask students to work in pairs on *Parts 1, 2,* and *3.*

Bring the class together to share answers and reasoning.

In *Part 1: Less Than a Penny,* focus attention on the part of each decimal that is less than a cent.

How do you interpret this amount? What part of a cent is this digit? How would you relate the amount to a penny, a dime, or a dollar? To which amount is it closest, and how do you know?

For example, in \$0.113, the 3 has a value less than 1 cent, $\frac{3}{10}$ of a cent, or $\frac{3}{1000}$ of a dollar. \$0.113 is between 11 and 12 cents, but closer to 11 cents.

In *Part 2: Less than 1%,* focus on the value of 0.5% as a decimal or a fraction. Ask,

If 1% = 0.01, how would you write 1.5% as a decimal? (0.015)

In *Part 3: Picturing Thousandths,* focus on the visual relationships between, 1, 0.1, 0.01, and 0.001.

Activity 3: Expanded Notation

Give an overview of expanded notation with whole numbers before having students begin this activity with the *Thousandths Cards* at hand. Ask:

How would you write 24 in expanded notation? (If students struggle, explain that the goal is to make the place value of each numeral explicit. For example, the '2' in 24, is 2 tens, not simply 2. Then try another.)

How would you write 825 in expanded notation? 8.25?

Activity 4: Eight Questions

Refer students again to the *Thousandths Cards.* Students work in pairs, preferably with access to the Internet.

This is the first time students are asked to use percents over 100. Connect these percentages to fractions that are more than one whole (such as $3/2$).When you bring the class together to discuss responses, spend some time ensuring that the decimal and percent equivalents of $1/8$ and $2/3$ are understood (Problems 3, 4, and 7).

How did you count by eighths using decimals?

Summary Discussion

Reflect on the introduction to the lesson (*Student Book*, p. 103). Ask:

How many 0.001's are in 1? (1,000)

How many 0.01's are in 0.1? (100)

How many 0.001's are in 0.01 (10)

Direct students to the *Reflections* (*Student Book*, p. 213).
To summarize ideas that arise about thousandths, ask students to write one idea they learned or want to remember about the value of decimals beyond the hundredths place, and then to write a question they still have.

Practice

Three Decimal Places. p. 111
For practice interpreting decimals to the thousandths place.

Rounding to the Nearest Tenths and Hundredths, p. 113
For practice using the calculator to convert from fraction to decimal and percent form.

Splitting Tips, p. 115
For practice estimating with fractions of percentages.

Calculator Practice

Fraction-Decimal-Percent Conversion, p. 112
For practice using the calculator to convert from fraction to decimal and percent form. Before assigning this practice, ask students if they know how to convert a fraction to a decimal. If they don't, prompt for the operation that would result in the calculator producing .5 as the equivalent of $1/2$. Test this procedure on the calculator with $1/4$ and then $1/3$, $2/3$, and $3/100$.

Extension

Rate Hikes—Ouch!, p. 117
For practice with fractions of cents.

Test Practice

Test Practice, p. 119

Looking Closely

Observe whether students are able to

Compare thousandths to hundredths, tenths, and ones

The goal here is for students not only to recognize how to say, picture, and write decimal numbers, but also to grasp how they relate in magnitude to each other. Relationships that are ten or hundred times bigger or smaller allow numerate people to compare quantities using multiplicative reasoning (.70 is hundred times greater than .007). This grasp of relationships using place value lays a foundation for understanding increases expressed in percentages, e.g., 200% or 250% more.

Write decimals in expanded notation.

Do students choose the right fraction or decimal to reflect each place? If not, review with the place value chart. Model naming the amounts in tenths and hundredths to reinforce the relationships. $\frac{3}{10} = \frac{30}{100} = \frac{300}{1000}$. While this may seem like an academic exercise, it is another way to see the associative property at work. Remind students that this property gives them the power to break numbers into parts.

Round decimals in the thousandths to the nearest 1, 0.1, 0.01

Can students talk through their reasoning, explaining to which place they are rounding, why, and what numbers are informing their decisions? Keep the discussion focused on the context when possible (as in the runners' activity), so that the need for precision drives decisions about rounding.

Rationale

New technology has made it possible for people to examine minute particles. Researchers investigate the impact of nanograms, amounts the size of a billionth of a gram (written $\frac{1}{1000000000}$ or .000,000,001). At the same time, workers in STEM fields, including nurses, calculate doses based on fractions of liters and milligram concentrations. In several of the community college degree and certificate program curricula, complex applications of measurement are prominent (National Center for Education and the Economy, 2013). Decimals appear in spreadsheets, in algebraic expressions and equations, so while considered Level B and C (5th-8th grade) content in the College and Career Readiness Standards, applications are wide-ranging and far-reaching.

Math Background

Students might be familiar with written or spoken decimal amounts in the thousandths or smaller if they have participated in conversations in their communities about trace amounts of pollutants in the air and water. If students have a working knowledge of the metric system, draw on the relationships among units to reinforce the idea of units being 10 times larger or smaller than the number in the place to the right or left. Draw on working knowledge some students might have based on money, measurement or other application.

Facilitation

Acknowledge that students often get flummoxed by long strings of numbers on the right side of the decimal point. To anchor their understanding, keep reinforcing the relationship of 10 times larger or smaller; write the decimal number as a fraction.

Making the Lesson Easier

Reading decimal numbers in the hundredths and thousandths can quickly become a tongue twister, especially for students whose native language is not English. If pronunciation becomes an obstacle, decide as a class on a way to talk about these amounts. Establish a classroom culture of confirming amounts in writing so that you can be sure the language is not getting in the way of understanding.

FACILITATING LESSON 8

Adding and Subtracting Decimals

How do we combine, take away, compare, or find the difference?

Synopsis

This lesson focuses on addition and subtraction using decimals. Students draw upon their prior knowledge of adding and subtracting whole numbers, fraction and decimal equivalence, and decimal values, and use this knowledge to make sense of the two operations involving decimals.

1. The class makes sense of an error about decimal addition.
2. Student pairs make decisions about the reasoning in three scenarios, each depicting a common dilemma involving addition or subtraction of decimals.
3. Students solve two sets of addition and subtraction problems to arrive at a reasonable rule for the two operations with decimals.
4. Students apply their understanding of addition and subtraction with decimals to measure tolerances.
5. The class summarizes a rule for adding and subtracting decimals.

Objectives

- Build on the meaning of addition and subtraction operations with whole numbers, extending understanding to addition and subtraction with decimals
- Use place value to judge the soundness of answers to addition and subtraction problems involving fractions, decimals, and percents

Materials/Prep

- Calculators
- Fraction strips (Optional; for *Activity 1*, students may want to refer to the fraction strips with decimals written on them)

Heads Up!

This lesson assumes that students have familiarity and ease working with fractions, particularly benchmark fractions. They should also be able to make estimates with benchmark fractions to solve problems, have reliable methods of operations on whole numbers, be able to relate decimal amounts to benchmark fractions, and be able to visualize and compare decimals up through thousandths. To solve problems in this lesson, they will call on the commutative, distributive and identity properties.

Opening Discussion

Start by posing the following situation:

You're at the store buying a family-size container of peanut butter ($6.89) and gum ($0.98). You quickly estimate what you will have to pay. The cashier announces your total and you are surprised—it seems way too high: $16.69.

What do you think happened? Turn to your neighbor and share with each other what you think happened here.

As students are sharing, listen for any ideas you want to make sure get shared with the whole group. If it hasn't come up, ask if anyone made a rough estimate.

Why is it an error to add the 6 in the price of the peanut butter to the 9 in the price of the gum?

As students explain about lining up the numbers according to decimal place, emphasize what they have already learned in the previous sessions: each decimal place has its own value.

End with,

Today we will work on two operations with decimals: addition and subtraction. We will look at how operating with decimals is related to what you already know: operating with whole numbers and with fractions.

Activity 1: Watch Out!

Explain to students that they will be working with a partner to reason about three scenarios. Encourage them to make drawings, or, for fractions, to use their fraction strips.

In Scenario 1, Anita is counting 2 volunteers' hours on garden workdays. One volunteer recorded mostly in fractions. One in decimals. What is each person's total and the total for both? Listen to students' reasoning. Encourage students to try converting fractions to decimals.

Scenario 2 focuses on the idea from the *Opening Discussion*, aligning digits for the addition of decimals. The correct total is 2.4 pounds. As students share this correct total, ask them what Elena might have done to get the 24 lb. instead.

Scenario 3 focuses on subtraction of decimals using the comparison model, which illustrates again the dilemma of how to line up decimals to subtract (or add).

As students finish and share what they found, say:

I saw some of you solving the problems using fraction strips. I need a volunteer to share how that was helpful.

In the second and third scenarios, I heard disagreement about the deli counter person's math and the reporter's math, respectively. I need a volunteer to explain this.

As students share and discuss, emphasize the importance of thinking about subtraction by using more than just the take-away model. Strategies like adding on are as valid as subtracting (take away and borrowing). Highlight the place value in both the adding up (going to the nearest multiple of ten) and subtracting back (regrouping).

End with,

What did you learn in this activity about adding and subtracting fractions and decimals?

Ask for specific examples. Then proceed to *Activity 2.*

Activity 2: Methods for Adding and Subtracting Decimals—They Have to Make Sense!

In this activity, students are presented with decimal addition and subtraction problems. They are asked explicitly why they line up the decimals and must explain the relationship between decimals and their fraction equivalents. They come up with a procedure for solving problems and those procedures are then shared, compared, and generalized by the group as a whole.

Start by reviewing the solutions to the problem from the *Opening Discussion* and that of the second scenario in *Activity 1.* Say:

In those two problems, you said it was important to line up the decimals to come up with the correct answer. I need a volunteer to read the numbers using tenths, hundredths and thousandths.

As students are saying the numbers (six and eighty-nine hundredths; ninety-eight hundredths; eight tenths; one and two-tenths; four tenths), make sure to also write them as fractions. Start with the first one, 6 $^{89}/_{100}$, and ask the class to help you with the rest.

Push them to explain why the correct answers, $7.87 and 2.4 lb., make sense when you look at these decimals in fraction form [for example, 0.98 is $\frac{98}{100}$; not 9 and $\frac{8}{10}$ as the cashier mistakenly used].

I'm hearing you say that place value—tenths, hundredths—is important when adding decimals.
Is that also the case for subtracting decimals? (Yes, equally important). **Explain.**

As students are sharing (yes, equally important, and giving their reasons), say:

You will next have an opportunity to solve several addition and subtraction problems involving decimals in the way that makes the most sense to you. Then you will develop a rule based on your method.

Direct students to *Activity 2* (*Student Book, Lesson 8*, p. 124). Explain that their task is to solve the problems however they prefer (including calculators), then to note patterns they see, and come up with a rule.

Set 1: Look for Patterns

a. $0.1 + 0.2 + 5.4 = 5.7$

b. $2.70 - 0.08 = 2.62$

c. $\frac{9}{10} - \frac{5}{10} = \frac{4}{10}$

d. $1.010 + 1.020 + 0.089 = 3.119$

Looking at the first one together, what strikes you about it—how are you seeing it? (They are all in tenths.)

Continue with the other problems in the set.

What patterns do you notice?

What rule could you write up that works for all the problems in the set? What is the main feature of your rule in terms of decimals (or fractions)?

Continue the discussion until there is clarity on adding or subtracting decimal numbers with like place values and fractions with common denominators.

Set 2: Keep Looking for Patterns—Develop a Method

a. $2.34 + 1.4 + 0.02 = 3.76$

b. $0.001 + 0.1 + 0.01 = 0.111$

c. $\frac{4}{5} + \frac{8}{10} = \frac{16}{10} = 1\,\frac{6}{10} = 1\,\frac{3}{5}$

d. $\frac{1}{4} + \frac{3}{8} = \frac{2}{8} + \frac{3}{8} = \frac{5}{8}$

Looking at the first one together, what strikes you about it—how are you seeing it? (One is in tenths and the others are in hundredths.)

Continue with the other problems in the set.

What patterns do you notice?

What rule could you write up that works for all the problems in the set? What is the main feature of your rule in terms of decimals (or fractions)?

Continue the discussion until there is again clarity on adding or subtracting decimal numbers with like place values and fractions with common denominators. You should end up with one rule that will apply to both sets.

Then ask students to use their rules to check the problems in Set 3 to determine which ones use the rules correctly.

Set 3: Correct or Not?

a. 1.002 + 14.01 = 15.012

b. 0.054 + 1.023 = 1.077

c. $\frac{2}{3} + \frac{1}{2} = \frac{3}{5}$

d. 1.45 + 2.104 = 2.249

End with a rule for adding and subtracting decimals that the class agrees with. The rule may specify lining up decimal points, or lining up same place values.

Activity 3: Tolerances—Give or Take

Begin by asking:

Does anyone know what tolerance is? What does it mean when used in a work context?

Then explain that tolerances are used especially in the manufacturing world to monitor limits above and below a certain standard, such as size or temperature.

Since tolerances often use very small limits, this is a good time to show a real-life context for the addition and subtraction of decimals.

Refer students to *Activity 3: Tolerances—Give or Take* (*Student Book*, p. 126). Have them work in pairs to address each of the situations.

When everyone is done, bring the class back together to discuss their results. Direct them to the *Vocabulary* (*Student Book*, p. 205) to define tolerance in their own words.

Summary Discussion

Ask students to turn to their reflections (*Student Book*, p. 213) and to write a paragraph about their understanding of adding and subtracting decimals. Students might start off by saying, "I understand …", or, "I still feel confused about …".

Practice

Where Is the Point?, p. 128
For practice with decimal addition and subtraction.

Closest Estimate, p. 130
For practice estimating sums and differences of decimals and fractions.

Calculator Practice

Fractions and Decimals, p. 131
For estimating addition and subtraction of decimals and then using the calculator to check their estimates and/or get an exact answer.

Extension

Gone Fishing, p. 132
For addition and subtraction in the thousandths. Students convert units in order to give reasonable answers in pounds and ounces.

Test Practice

Test Practice, p. 133

Looking Closely

Observe whether students are able to

Build on the meaning of addition and subtraction operations with whole numbers, extending understanding to addition and subtraction with decimals

If students automatically line up numbers in a column and add or subtract without thinking about place value, you will need to intervene. They might do better aurally and with mental math. If they get problems right, see if they can replicate what they did mentally and intuitively, keeping ones, tens, and possibly tenths and hundredths together.

Students struggling with addition of decimals could benefit from working on *Practice: Where Is the Point?* (*Student Book*, p. 128). Students needing practice converting fractions to decimals before combining them could benefit from *Calculator Practice: Fractions and Decimals* (*Student Book*, p. 131).

When solving subtraction problems, do students know that adding on (instead of subtracting) is a reliable strategy? If students are uncertain of their answers, encourage them to use a number line to demonstrate the distance between two numbers or to make comparisons.

Use place value to judge the soundness of answers to addition and subtraction problems involving fractions, decimals, and percents

Do students recognize answers that don't make sense? Students' ability to judge the reasonableness of the solutions depends on their understanding of the situation, number sense, operation sense, and place value. If students seem to struggle, ask them to explain the problem in their own words or to show you a picture of what they understand. Check for understanding of relative size. Which quantities are much more or less than $10 or more than a tenth of a second? Students have many opportunities in this lesson to pay attention to implications of the decimal point in adding and subtracting with multi-digit numbers.

Rationale

Addition of fractions, decimals, and percents is routine in daily life. Solving problems with a calculator is practical and a valid strategy, yet it is important for students to understand a problem well enough to make a reasonable estimate and to defend their solution. The problems in this lesson exemplify the dilemmas that crop up when adding and subtracting decimals.

Math Background

Combining quantities to get a larger amount makes sense with whole numbers, but with fractions and decimals, the answer is often not obvious at first glance.

Adding decimals demands close attention. When adding 1,206 + 3.02, for example, if the numbers are aligned left to right and the decimal point is ignored, the answer will seem to be 1,508. If the place value of numbers is well understood, the correct answer is easy to find; the first number is 1,206, and if we add 3.02 to that, aligning the decimal point, we end up with 1,209.02. Problems with decimals are a matter of language. Understanding that the decimal point is used to separate the whole number portion of the number from the fractional portion of the number will ease the confusion.

Facilitation

Making the Lesson Easier

Do additional "Where Is the Point?"-type questions with the missing point in the answer before moving to those in which the point is missing from a number in the question.

Making the Lesson Harder

Assign *Extension: Gone Fishing* (*Student Book*, p. 132). Ask how students got the incorrect numbers, in addition to explaining the correct answer.

LESSON 8 IN ACTION

One educator told a puzzling story to introduce Activity 3: Tolerances—Give or Take.

> An educator explained to her peers:
>
> "I ran out of gas and as I was pushing my car to the pump, my friend with the same car, made in the same year, also was pushing her car into the gas station. We were both completely out of gas, but when I filled up and when she filled up, we had to pay different amounts. Why?"
>
> The group talked about all kinds of reasons and eventually about how manufactured parts might be slightly different from each other. One gas tank could be the slightest bit larger or smaller than another. A 12-gallon tank could be within + or – 0.25 of 12 gallons. This idea is called tolerance and many manufacturing companies keep track of tolerance using decimal measurements.
>
> *New York City Educators*
> *New York*

FACILITATING LESSON **9**

Multiplying Decimals

What is 2.5 times this amount?

Synopsis

Students draw upon their knowledge of whole number and fraction multiplication as they examine decimal multiplication. They expand their knowledge as they consider that decimal multiplication can make things smaller, just as it does with multiplication of fractions.

They solve multiplication problems involving numbers less than 1 and also greater than 1. They explore different ways to multiply decimals, including the U.S. standard algorithm.

1. Students visually show what multiplication of decimals looks like.
2. Students compare multiplication of decimals to multiplication of fractions and move to explain the U.S. standard algorithm.
3. Students find errors based on faulty calculator use.
4. Students explore shortcuts to multiplying with decimals.

Objectives

- Connect understanding of multiplication with whole numbers and fractions to multiplication with decimal numbers
- Decide upon a reliable method for multiplication with decimal numbers
- Use visual models and patterns to extend understanding of multiplication short-cuts with whole numbers and decimal numbers
- Apply the properties of arithmetic (e.g., commutative, distributive, associative) to decimals

Materials/Prep

- Items for counting such as paper clips, colored chips, or pennies (about 50 per student)
- Fraction strips (from earlier lessons), graph paper, rulers
- *Blackline Master 10: 100-Block Grids*, one or two per small group
- Easel paper, markers

Opening Discussion

Get students thinking about multiplication of whole numbers. Distribute items for counting to each student and ask:

How could you explain to your child (or a friend's child) what 3 × 12 means?

Ask students to think about this without using paper and pencils and to share with a partner before they share with the class. Encourage them to use objects to show their reasoning. Their explanations will likely include 3 added 12 times—12 groups of 3 pennies. Or, they could show 12 added 3 times—3 groups of 12 pennies. If nobody mentions that 3 × 12 is the same as 12 × 3, bring it up yourself. Ask:

How are the two problems, 3 × 12 and 12 × 3, related? How are they the same? How are they different?

Acknowledge that the groups (3 groups of 12 chips and 12 groups of 3 chips) are different, but that the total number of chips is the same. Ask:

How would you explain what 12 × 3.5 means?

Again, tell students to think about this without using paper and pencils, and to share with a partner before they share with the class. Encourage them to use objects to show their thinking but stand back while students puzzle this out.

Have students share their reasoning. Their explanations will likely include 3.5 groups of 12, breaking apart the 3.5 into 3 + 0.5 and multiplying each by 12: 3 × 12 plus 0.5 group of 12: (3 × 12) + (0.5 × 12) = 36 + 6 = 42. They used the distributive property. If nobody mentions that 3.5 × 12 yields the same result as 12 × 3.5, bring it up yourself.

How are the two problems related? How are they the same? How are they different?

Acknowledge that the groups (3.5 groups of 12 chips, and 12 groups of 3.5 chips) are different, but that the total number of chips is the same.

Students might make a sketch of both situations that looks something like this:

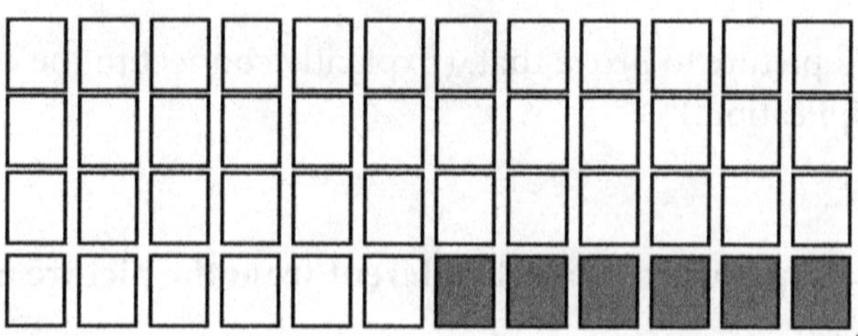

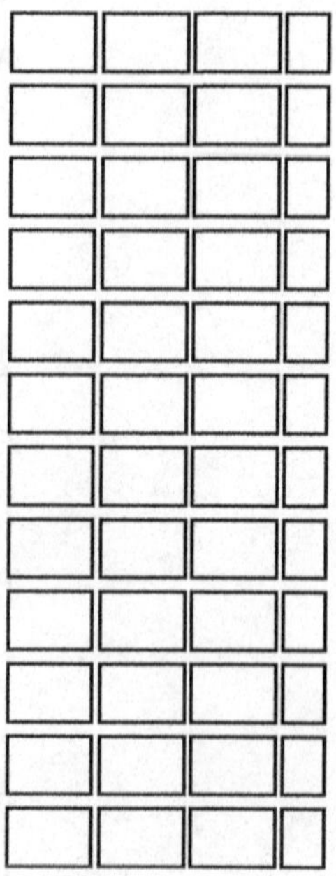

The arrays demonstrate the two ways that the same problem can be solved. In both cases, the distributive property is used: in the first, we are looking at groups of 12. There are 3 groups of 12 (36) and another half row of 12 (6). In the second, there are 12 groups of 3 (36) and 12 groups of 1/2 (6); each yielding a product of 42. In both cases, the products of the parts are added to get the total product: $36 + 6 = 42$.

So $12 \times 3.5 = (12 \times 3) + (12 \times 0.5) = 42$ and $3.5 \times 12 = (3 \times 12) + (0.5 \times 12) = 42$. Tell students that in this lesson, they will think about these ideas as they solve multiplication problems involving fractions and decimals.

Activity 1: Show Me ...

This activity gives students the opportunity to extend to decimals their understanding of the meaning of the multiplication operation.

Bring students' attention to *Activity 1: Show Me...* (*Student Book*, p. 136). Distribute or make available virtual 100-grids, rulers, and graph paper. Students may want to use their fraction strips. Have people work in teams on the visual representations and situations connected to each of three problems. Then bring the class together to publicly share representations and real life situations.

Facilitate the discussion for each problem, mentioning the following points if they do not come up.

1. Show me 3×0.1

Do some people interpret it as three one-tenths? ($\frac{1}{10} + \frac{1}{10} + \frac{1}{10}$)

Do others interpret it as $\frac{1}{10}$ of 3?

If both interpretations do not come up, ask:

Is 3 × 0.1 the same as 0.1 × 3?

Ask for a picture to prove that. (Explicitly connect to the commutative property of multiplication.)

Then ask,

How is the picture of 3 × 0.1 different from the picture of 0.1 × 3?

2. Show me 1.75 × 10

Do some people interpret it as one and three quarters of ten?

Do others interpret it as ten sets of one and three-quarters? (1.75 + 1.75 + …)?

If both interpretations do not come up, ask:

Do these two interpretations yield the same answer?

Ask for a picture to prove that. Explicitly connect to the commutative property of multiplication. Ask:

What mathematical principle does 1.75 × 10 = 10 × 1.75 show?

Look to see whether anyone has interpreted 1.75 as 10 + ¾ of 10 = 17.5 — a good application of the distributive property. If not, write the following on the board, asking how this connects to anyone's picture:

1.75 × 10 = 1 × 10 + 0.75 × 10 = 10 + 7.5

3. Show me 12.5 × 0.5

Do some interpret it as twelve and a half halves? Do others interpret it as a half of 12 and a half?

If it does not come up, ask if the two interpretations yield the same answer. Ask for a picture to prove that. Once again explicitly connect to the commutative property of multiplication.

Summarize by asking:

How did you use what you know about the meaning of multiplication to understand these problems?

How did you use your knowledge of fraction multiplication?

Activity 2: Methods for Multiplying Decimals—They Have to Make Sense!

In this activity, students are presented with sets of decimal multiplication problems. They solve these problems using fractions, visuals, calculators, and whatever algorithm they are comfortable with for multiplying decimals. Say:

You have made sense of what some decimal multiplication problems mean with pictures and situations. Now, turn your attention to mathematical steps —called algorithms—for multiplying decimals.

Direct students to *Activity 2: Methods for Multiplying Decimals—They Have to Make Sense!* (*Student Book,* p. 138). Explain that they will now look at different ways to solve decimal multiplication problems and will develop their own rules for doing so.

Review the first problem of *Part 1* together:

$3 \times 0.1 = ?$

Step 1—Let's try it together, working it out in fractions. What is $3 \times \frac{1}{10}$? ($3 \times \frac{1}{10} = \frac{3}{10}$)

Translate that into decimals for Step 2? ($3 \times 0.1 = 0.3$)

Now Step 3, check it on a calculator.

Step 4, show how it makes sense that $3 \times 0.1 = 0.3$.

Now look at the other two problems. Work with a partner to state the steps for solving problems that look like these examples.

Once students have completed the worksheet for *Part 1,* convene the group and ask:

What do all the problems have in common in Set 1?

Continue the discussion until there is clarity and agreement on one or more sets of steps for multiplying a whole number times a number with a decimal amount in the tenths, hundredths, or thousandths.

Direct students to *Part 2.* Remind them to picture the problems before writing the steps for solving the problems and comparing them with a partner's list. When pairs are ready to talk, review their responses:

What do all the problems have in common in Set 2?

Continue the discussion until there is clarity on multiplying two numbers, both with decimal amounts. Once volunteers share the steps, ask:

What happens to the amounts less than one?

If you had to talk about these amounts as fractions, what happens to the number of pieces and the size of the whole?

The take-home point here is that multiplying by a decimal gives you a fraction of what you started with. To be precise about how big that fraction is, you may need to divide the unit into smaller pieces. If whole numbers with decimal amounts (4.8×90.6 for example) are being multiplied, the whole number in the product increases. At the same time, the decimals to the tenths or hundredths, or thousandths imply that you are paying attention to the number of pieces out of 10 or 100 or 1,000 or even smaller amounts.

Direct students to work on Set 3. When they have found commonalities, listed steps, and tested their methods, convene the group. Listen for insights about the results when two numbers with decimal parts are multiplied. As always, if students can predict a ballpark answer, they can eyeball their answers, making sure they are on the right track. Reinforce ideas from the discussion of Set 2, by asking:

What do all the problems have in common in Set 3?

What happens to the amounts less than one? Why is that?

If you had to talk about these amounts as fractions, what happens to the number of pieces and the size of the whole? How could you predict how precise the answer will be? (to what place?)

In asking these questions listen for the idea that the place value of the answer is connected to the number of places in the numbers being multiplied. For example, tenths by tenths makes for pieces out of 100, so the answer is in the hundredths. Tenths by hundredths is thousandths, so the answer is in the thousandths.

Activity 3: Calculators—What Happened?

Explain to students that most adults tend to use a calculator to solve complicated decimal multiplication problems. Even though calculators are extremely useful for doing the calculations, it is important that students know if the answer makes sense, especially since it is very easy to incorrectly punch in numbers or symbols on a calculator.

Refer students to *Activity 3: Calculators—What Happened?* (*Student Book,* p. 142) and ask them to work with a partner to figure out what could have happened to lead to an incorrect answer.

As you observe students working, ask if they can use what they know from their examination of the sets of problems in *Activity 2* to predict how many places the answer should have.

Did the size of any of the answers (either your predicted answers or the calculator's answers) surprise you?

Students might say that when you multiply by a small number, e.g., a number in the thousandths, the effect is strong. Multiplying 7.4 by the decimal 0.002 results in a product that is much less than one. Before moving on, you might ask students to assess their confidence in their intuition.

What are you noticing about the effect multiplication with decimals will have?

Activity 4: Shortcuts—Multiples of 10

Students have done some exploration of multiplying by 10, 100, and 1,000 in earlier lessons and in other *EMPower* books. In this activity, they will look at tenths and hundredths as well as tens and hundreds.

Refer students to *Activity 4: Shortcuts—Multiples of 10* (*Student Book*, p. 143). Have them work in pairs. When everyone is done, bring them back together and ask them to describe the shortcuts that they discovered. You might want to capture these on easel paper and post for referencing later. Then ask:

Predict a shortcut for multiplying by 1,000.

What if you were to multiply a number by 0.001?

Describe a shortcut for multiplying with any multiple of 10

How can the patterns you observed help you predict the answers to other problems with decimal multiplication?

If students aren't ready to answer this problem, leave it for now. They can continue to examine multiplication with decimal numbers and they may have more to say after the two math inspections.

Math Inspection: Exponents

This inspection is meant to help students multiply numbers with both positive and negative exponents. By looking for patterns, they begin to develop a solid base for understanding important concepts such as what exponents are and how they behave.

Ask students to complete Part 1. The focus is primarily to see/understand what multiplying a number by itself a number of times means, and how to write it using exponents. All of the exponents in this table are 2, as each number is multiplied by itself two times.

In Part 2, Set 1, they are examining powers of 10 and seeing a pattern (that the exponent indicates the number of 0s attached to the 1), as well as what a negative exponent means. Be sure to stress that a number (other than 0) to the 0 power is 1. If they create a pattern, for example,

$$10^3 = 1{,}000,$$
$$10^2 = 100,$$
$$10^1 = 10,$$
$$10^0 = 1,$$
$$10^{-1} = 0.1,$$

etc., they can begin to see definite patterns without having to memorize rules.

In Set 2, they are exploring powers of 2, going down from 2^4 to 2^{-3}. Again, they are exploring what a negative exponent means and how to represent it numerically. In Set 3 they explore powers of 1 to find a pattern, and in Set 4 they give their own base and find the result with different exponents.

In Part 3, they operate with exponents, finding products as well as quotients. They write the numbers in expanded form to make the problem more visual and work from expanded to exponential form as well. In the end, they will notice that when multiplying two numbers of the same base and with exponents, the product is that base to the sum of the exponents. If they are dividing numbers with the same base, the quotient is that base to the difference between the exponents.

Math Inspection: Break Apart the Numbers

This inspection is meant to help students make sense of the U.S. standard algorithm for multiplication by connecting it to what they already know about breaking numbers apart as they use the distributive property to find the product of decimal numbers.

Math inspections will have more impact if students are given time to notice patterns, come to generalizations, and justify their reasoning. This will provide them with a solid base for understanding what distributive means.

Ask students to explain how the two solutions to the first problem are the same, and what is going on in the strategy that Dana used when she broke the numbers apart. Specifically, ask why it is that you can do this.

As they solve the subsequent problems, ask them if they would be able to solve any of these mentally, without paper and pencil, and give them one to try.

Math Inspection: Shortcuts with 0.5 and 0.25

This inspection is meant to help students find more efficient ways to multiply by 0.5 and 0.25. As they solve groups of problems involving ½ and .5, multiplication and division, they notice patterns that allow them to make a generalization about shortcuts for multiplying by 0.5 and 0.25, e.g., multiplying by 0.5 and dividing by 2 are the same; multiplying by 0.25 and dividing by 4 have the same outcome.

Ask students to solve the six problems in each set and compare the results. Then ask them to compare similar problems: were they all multiplication? Were they multiplying by the same number? If not, find what was the same by pairs (for example, in Set 1, **0.5 × 326** and **½ × 0.326** and then **0.326 ÷ 2**).

In Set 2, after they solve each problem, ask students what they notice about the product or quotient, and ask them why they think the answers are what they are. For example, **0.125 × 8 = 1**, so ask, "What's another name for 0.125? How do you know?"

Summary Discussion

Take time to summarize and generalize about the effect of multiplication:

Given what you now know about decimal multiplication, decide whether each of these statements is correct:

1. When you multiply two decimal numbers, the result (product) is always smaller than the either decimal. (Not always true, but in certain cases it is.)
2. When you multiply by 100, you move the decimal one place to the right. (False)
3. When you multiply by 0.1, you move the decimal one place to the left. (True)

Ask students to reframe the false statements or to provide examples to support their positions. You might hear statements such as this for #2:

> "When you multiply by 100, the number will be two places larger than it started. Add two zeroes, which pushes the decimal point to the right two places."

Direct students to the *Vocabulary* (*Student Book*, p. 205), and *Reflections* (*Student Book*, p. 214) to record terms and ideas they want to remember.

Practice

Draw It!, p. 153
For practice finding a visual way to represent multiplication of decimals.

Estimate Answers to Decimal Multiplication, p. 155
For practice with estimating decimal products and determining where to put the missing decimal points.

Squaring Fractions and Decimals, p. 158
For practice using exponents to write fractions and decimals.

Using an Area Model for Multiplication, p. 159
Multiplying decimals using the rectangle array, or area model, as a visual.

Calculator Practice

Decimals, Decimals, p. 162
For practice using the calculator to multiply decimals and look for patterns.

Test Practice

Test Practice, p. 163

Looking Closely

Observe whether students are able to

Connect understanding of multiplication with whole numbers and fractions to multiplication with decimal numbers

At this point students may have a visual or intuitive sense that enables them to imagine the action of multiplication. A unit or quantity repeatedly added can work as a visual. Extending this idea to fractions and decimals means imagining that initial quantity multiplied by an amount less than one and visualizing that, even accumulated, the total might not be more than one. Visual models, sketches, grids, or arrays can make this apparent. Multiplication with decimals, like fractions, may not result in a product that is larger than the multiplicands.

Decide upon a reliable procedure for multiplication with decimal numbers

How are students handling numbers with so many digits? Are they paying attention to the placement of the decimal point? Probe for explanations that make sense both for carrying out the multiplication and for checking the answer is the correct order of magnitude. Students might think it is strange that you can multiply a whole number by a whole number and get a whole number, but if you multiply tenths by tenths you get hundredths. The point of arriving at a reliable procedure is not to spend lots of time doing multiplication problems by hand, but rather to examine problems and observe patterns so that students build an intuitive sense of the magnitude of their answers.

Use visual models and patterns to extend understanding of multiplication short-cuts with whole numbers and decimal numbers

With a strong sense of place value, students can use shortcuts to find products for problems like .37 × 10 or .02 × 200. Challenge students to explain how multiplying is affecting the quantities. "Moving the decimal point" does not adequately describe the fact that the quantity is increasing or decreasing. The idea of place value shifting to create a smaller or larger number gets closer to explaining multiplication's effect. Probe for further explanation.

Apply the properties of arithmetic (e.g., commutative, distributive, associative) to decimals

Decimals are similar to whole numbers in that one or more in a multiplication problem can be broken apart and the product will be the same as long as each sub-part is multiplied. So 0.24 × 1.5 can be thought of as 0.2 × 1.5 + 0.04 × 1.5. As is the case with whole numbers, multiplication with decimals is commutative. 1.5 × 0.24 is equivalent to the expression 0.24 × 1.5. They can be solved using the distributive property:

$$0.24 \times 1.5 = (0.2 \times 1.5) + (0.04 \times 1.5) \text{ and}$$

$$1.5 \times 0.24 = (1 \times 0.24) + (.5 \times 0.24).$$

Do students see that they have options to simplify the problems so that they can come up with an estimate, check their calculations, or justify their answer?

Rationale

Sometimes repeated addition is a good model for multiplication. However, problems such as ¾ × ½ are more difficult to understand with repeated addition. Contextualizing it—for example, a ¾ ft. by ½ ft. rectangle, inside of which you want to put a photograph—gives meaning to the operation. You can almost see the frame in which you want to insert the photograph. This lesson focuses on visualizing the operation, solving problems, and having tools—mental, visual, and computational—to solve multiplication problems.

Math Background

Not all the ideas about multiplication with whole numbers translate to fractions and decimals. That multiplication is commutative and that multiplication by 1 results in the original number are both true for all numbers. That multiplication results in a larger number than the original, however, does not hold for numbers less than 1. In fact, those products are smaller (e.g., 3 × ⅙ = ½, whereas 3 × 6 = 18). It is also true for all numbers that you can divide the number by itself or multiply it by its reciprocal to get 1. However, this is not as evident with decimals (e.g., ¾ × ⁴⁄₃ = 1 is the same as ¾ ÷ ¾).

The commonalities and differences are important, and the ideas that hold for whole numbers are useful in trying to make sense of the ideas about multiplication involving fractions and decimals. Although in some cases converting a fraction to a decimal might be easier, making an estimate first continues to be important. For example, .6 is approximately a half, so when multiplying it by another number, say 3, estimate the answer to be about ½ of 3, or 1½. When multiplying by a number larger than 1, say 2.9, an estimate is also useful. For example, in multiplying 96 × 2.9, you can estimate that 2.9 is about 3. A number of strategies could be used to find 96 × 3 relatively easily. Knowing the result of that estimate—for instance, it must be less than 300—could be very helpful.

Context

The problems in these activities focus attention on the structure of number and the meaning of algorithms. You might need to supply contexts so your students can apply their ideas in a familiar, real-world scenario and use the results to gauge the value of their ideas. The context of sewing, woodworking, liquids, and concentrations all might be familiar. Have examples on hand, such as a meter stick and a half-cup measure.

Facilitation

This lesson will take more time than most. Save easel paper and review. Take a picture and re-create the display if the group's thinking was recorded on an erasable surface.

Making the Lesson Easier

Probe for or provide context when problems do not provide any (Activity 2, for example). Provide frequent check-ins while students are working on Math Inspections.

Bring in the language of "groups of" and "sets of" instead of reading multiplication as "times."

Ask students if they can solve a multiplication problem of their choice using the distributive property.

Making the Lesson Harder

Ask students to pick problems with no context and to create a story problem to match them.

FACILITATING LESSON **10**

Dividing Decimals

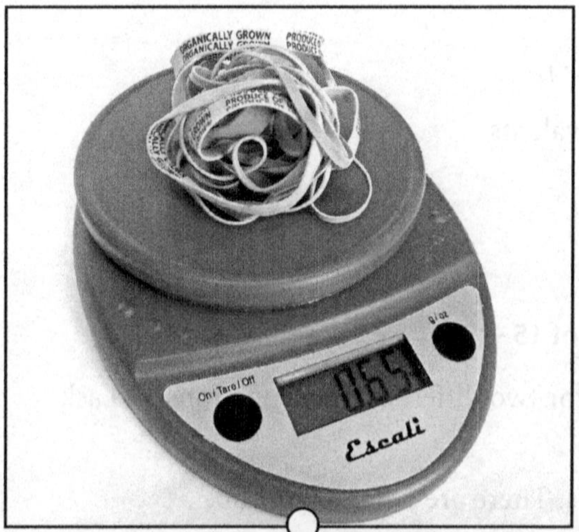

How do you think about division?

Synopsis

If students have studied division before, they may be aware that problem solvers can think of division either as 1) sharing, splitting, or partitioning an amount or 2) finding how many groups of a specific size or quantity are in an amount (How many 8's in 64?). In this lesson, students will review the two ways to look at division with decimal numbers. They will also explore why the traditional algorithm of the moving decimal point works.

1. Student pairs interpret and make a depiction of a numbers-only division problem, then create a real-life story to determine the answer.
2. Students share their remembered paper-and-pencil procedures for dividing decimals, then explore why the algorithms work.
3. Given a monthly expenditure, students calculate the weekly expenditure, explaining why division by 4.3 makes sense.

Objectives

- Interpret division with decimals as an act of splitting an amount or finding how many groups can "fit into" an amount, extending these ideas from the study of division with whole numbers
- Match verbal language and symbolic notation for division to a concrete model
- Compare and contrast *a*/*b* with *b*/*a* in various notation

Materials/Prep

- **IMPORTANT:** Read the *Rationale* and *Math Background* in *Lesson 10: Commentary*, pp. 119-120, before preparing this lesson
- Calculators
- Rulers
- *Blackline Master 10: 100-Block Grids*
- Fraction strips marked with equivalents

Opening Discussion

Open with the following question:

How would you explain the meaning of 15 ÷ 5?

Ask students to draw 15 ÷ 5. Looking for two different interpretations and ask volunteers to draw them on the board:

It means "split 15 into 5 groups. There are 3 in each group."

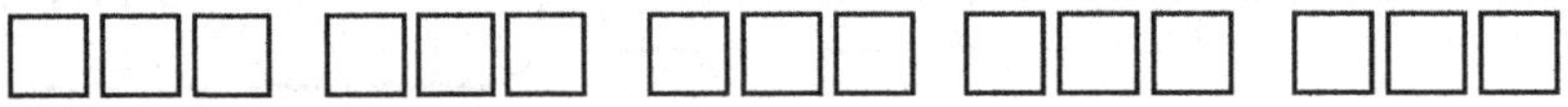

It means "how many 5's are in 15?"

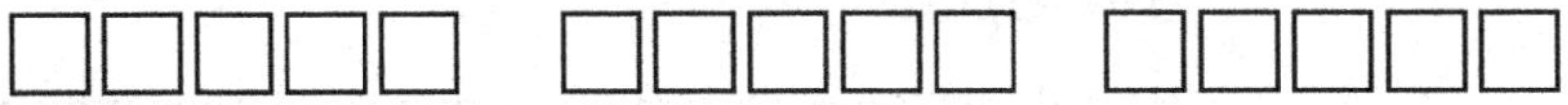

If students do not offer both interpretations, make them explicit.

Summarize:

When you see division, you can use either way to make sense of the problem.

Which way helps you make sense of these equations? Which does not?

Write the following four problems on the board or display them digitally, and ask students to explain the meaning, either by using the idea of splitting or by asking "how many ___ in ___?"

1. 4.6 ÷ 4.6 =
2. 10 ÷ 2.5 =
3. 0.25 ÷ 2 =
4. 2.5 ÷ 10 =

Invite discussion and ask students to make diagrams on the board. The first two problems are most likely interpreted as "how many ___ in ___?" The last two are more easily seen as splitting.

Say:

Let's draw a division problem.

Post the division problem 22 ÷ 5 = ___.

Solicit one volunteer to draw a picture of the problem, and ask the class whether they agree with the representation. This example is a partitioning situation, where the diagram shows five groups of $4\frac{2}{5}$ each.

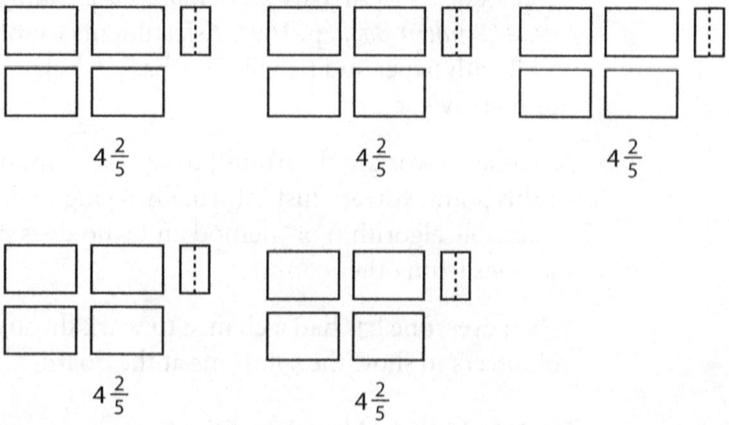

Ask students for another way to interpret the problem. Students may show four whole groups of 5, and 2 left over, as in this example.

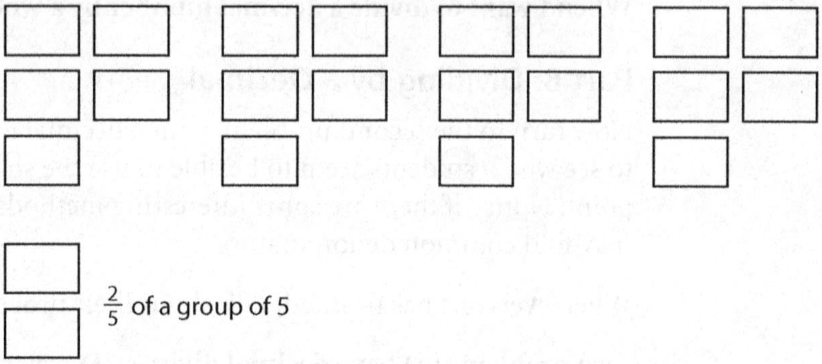

If students need another example, direct them to their lesson introduction and ask them to draw the two contextualized interpretations of $12\frac{1}{2} \div 5$.

Keep both interpretations in mind as you work on the problems in this lesson.

Hand out rulers, copies of *Blackline Master 10: 100-Block Grids* and have fraction strips handy as thinking tools.

Activity 1: What Is the Message?

Set up the activity by saying:

You know that math symbols always mean something. You can show the meaning with a picture or diagram, or connect the math to a real-life situation.

Direct students to *Activity 1: What Is the Message?* (*Student Book,* p. 166). When students are done, have them share their diagrams publicly. Watch out for order ($12.5 \div 100 \neq 100 \div 12.5$) and for understanding the difference between $3.25 \div 5$ and $3.25 \div 0.5$.

Activity 2: Methods for Dividing Decimals—They Have to Make Sense!

Refer students to *Activity 2: Methods for Dividing Decimals—They Have to Make Sense!* (*Student Book*, p. 169). Ask students to solve the two problems in Parts A and B with paper and pencil. One has a whole number divisor, the other has a decimal divisor.

As students work, walk around to see the computational strategies they use. At this point, you are just informally trying to determine who already uses a reasonable algorithm or method and who does not. Notice if there are some methods from other countries.

When everyone has had a chance to work through the problems, ask for volunteers to show the solutions at the board.

Part A: Whole Number Divisors

The first problem has a whole number divisor. Develop a method or a rule for dividing by a whole number:

When I want to divide a decimal number by a whole number, I ______________.

Part B: Dividing by a Decimal

Now turn to the second problem, with a decimal divisor. Pay special attention to see which students seem to be able to use the short-cut of moving the decimal point. Notice if there are some interesting methods from other countries. Some may find common denominators.

When everyone has finished, talk about their procedures.

Two problems (f.) have decimal divisors. Develop a method or rule for dividing by a decimal:

When I want to divide by a decimal number, I ________________.

Say:

Let's explore *why* these rules work.

Heads Up!

You may see two algorithms—procedures or methods—that people have been taught for paper-and-pencil decimal division. Even though these procedures appear different, they are based on the same principle: that is, equivalent ratios yield the same answer—or quotient.

METHOD 1

Some people *move* the *decimal point* in the divisor and then again in the quotient the same number of places, like this:

$.2\overline{)8} \longrightarrow 2.\overline{)80}$

Why does that make sense?

($^{80}/_{2}$ is equivalent to $^{8}/_{0.2}$ because $^{8}/_{.2} \times (^{10}/_{10}) = {}^{80}/_{2} = 40$, and multiplying by $^{10}/_{10}$ is the same as multiplying by 1.)

METHOD 2

Some people *find common denominators* for the divisor and dividend, and then find it easier to think about what the problem means, like this:

$$^{8}/_{0.2} \longrightarrow {}^{80}/_{10} \div {}^{2}/_{10} = 40$$

($^{80}/_{2}$ is equivalent to $^{8}/_{0.2}$ because $^{8}/_{0.2} \times (^{10}/_{10}) = {}^{80}/_{2}$, and multiplying by $^{10}/_{10}$ is the same as multiplying by 1.)

Ask students:

Which one of these methods is closest to your procedure?

Activity 3: Weekly Expenses

Refer students to *Activity 3: Weekly Expenses* (*Student Book*, p. 171).

Ask pairs of students to work together to think about why the bank instructions say to divide the monthly expense by 4.3. Share reasons.

Then ask students to work individually on the next part, using Christina's spreadsheet.

Given the monthly amounts for a year's expenses, the students' task is first to find the average monthly expenses, and then to estimate and to compute the exact weekly amounts to be reported on the form. Encourage calculator use to arrive at exact calculations.

After students have finished, ask them to compare answers with their partners and to share methods they used to do the mental estimations and the exact calculations. Then bring the class together and reinforce the variety of useable methods:

Who had an estimation method that worked well?

Spend time on the mental math methods. Ask individuals to come up to the board to explain and record their mental math strategies. Some may choose to reason with multiplication, others with division. There are many valid ways to do this. For example, some students might reason about splitting up the money this way:

- **Rent:** Because it is the same every month, $1,000, a good estimate is $200 to $250 per week. $250 would be a little too much and $200, a little too small.
- **Cell Phone:** The bill each month is between $67 and $72. The average each month is about $70, or $15–20 per week.
- **Heat:** Some months, there was no bill. Over the year, the cost was $1,400. If you spread that over 52 weeks, the weekly cost would be somewhere around $26.

- **Electricity:** This hovers around $40 per month, or somewhat less than $10 per week.

Acknowledge any paper-and-pencil methods. Depending on where and when they went to school, students may have learned different algorithms. Ask them to show those.

Who has a paper-and-pencil method that works well?

When you used the calculator, what did you have to keep in mind? When you estimated weekly expenses, did you think about the sharing model (or splitting or partitioning), or did you think about how many groups of _____ in _____?

Listens to student reasoning with examples from the categories of rent, cell phone, etc. Listen for the idea that order makes a difference when keying in the numbers. Summarize by remarking on the variety of ways a number can be divided or split.

Summary Discussion

Ask volunteers to share what they learned in this lesson about decimal division.

What is clear? What remains unclear?

How does working with decimal numbers in division seem the same or different from working with whole numbers?

Direct students to *Vocabulary* (*Student Book*, p. 206) and *Reflections* (*Student Book*, pp. 214-215), where students record what they want to remember.

Practice

Four Ways to Write Division, p. 173
For practice writing division problems using different symbols.

Target Practice: 0.1, 0.01, 100, p. 174
For practice using calculators to try to get close to the target number.

Which Is Not the Same?, p. 176
For looking at different ways to write division problems.

Where's the Point?, p. 178
For using reasoning and estimation to determine decimal point placement.

Multiplication and Division Patterns, p. 179
For practice looking for patterns with multiplication and division of decimals, then checking with calculators

Division Patterns, p. 180
For practice looking for patterns with simple division problems.

Think Metric, p. 181
For practice using the metric system to divide with decimals.

Free Choice, p. 185
For extra practice calculating mentally, with calculators, or on paper.

Calculator Practice

Decimal Division, p. 182
For practice using the calculator to determine answers to division with decimals.

Way Under Average?, p. 183
For practice calculating averages and comparing to a new piece of data.

Extension

Geometric Formulas, p. 186
For applying understanding of decimals to geometric formulas.

Test Practice

Test Practice, p. 188

Looking Closely

Observe whether students are able to

Interpret division with decimals as an act of splitting an amount or finding how many groups can "fit into" an amount, extending these ideas from the study of division with whole numbers

Are students secure with the partitioning model for division? If students are unsure of division as an act of splitting, use manipulatives. Also, refer to the idea of unit price. If you know 5.4 lbs. of hamburger cost $11.89, what is the cost of one pound?

When faced with a situation of dividing $500 by 4.3, are students able to anticipate that the answer must be somewhere between $100 and $125? That is an important idea to develop in this lesson. As far as arriving at the exact answer, encourage a variety of strategies. A calculator is often most efficient, but for those students who prefer a paper-and-pencil procedure, that is fine as well. If you see students using the distributive property to simplify the problem, bring this to the awareness of the whole class. For example, $500 \div 4.3 = (430 \div 4.3) + (70 \div 4.3) = 100 + 16.27 = 116.27$.

Note: to use the distributive property when dividing, only the dividend can be decomposed. That is, $500 \div 4.3$ could not be solved with $(500 \div 4) + (500 \div 0.3)$ because the important unit here is the group size, which is 4.3.

To cement the connection between division and splitting the monthly payment into weekly amounts, use a calendar. The monthly amount must be split proportionately among the four-plus weeks.

Spiral back to multiplication strategies. Though this lesson centers on ways to divide, the contexts can also be used to reinforce multiplication strategies by asking students to use multiplication to check their work (in addition to estimating the answers).

For more practice estimating the result of dividing an integer by a mixed number, assign *Practice: Where's the Point?* (*Student Book*, p. 178) and *Calculator Practice: Decimal Division* (*Student Book*, p. 182).

Match verbal language and symbolic notation for division to a concrete model

Are students consistently making accurate connections between the notation, a diagram or sketch, and words? This is one of the most important goals of the lesson—to see that the notation has meaning. Give students opportunities throughout the lesson to demonstrate understanding of division with decimals in various ways—in words, drawings, and with manipulatives.

As a warm-up, consider preceding *Activity 1: What's the Message?* with additional sketches of different division problems. Are students able to visualize the splitting of a fraction or decimal number into a number of equal parts? The abilities to make meaning of the notation, visualize it concretely, and apply it to a situation are the bedrock for anticipating the approximate answer. Again, allow for various procedures to arrive at an exact answer, either using a calculator or a paper-and-pencil method that students might have learned previously.

Compare a/b with b/a in various notations

Are students secure with the order in division notation? During the *Opening Discussion*, emphasize the need for common understanding when someone writes or speaks about division. If, for instance, someone reads a problem as "5 divided into 7½," he or she is commonly understood to mean, or 7.5 ÷ 5. A common mistake would be to write 5 ÷ 7½.

Which number belongs inside the box when writing out a division problem? Which number is entered into the calculator first? Although such issues are less likely to arise when solving contextualized problems, students often encounter problems translating division expressions during testing situations. If students need practice with equivalent notation, assign the practices *Four Ways to Write Division* (*Student Book*, p. 173) and *Which Is Not the Same?* (*Student Book*, p. 176).

Other than noticing that the expressions a/b and b/a are not equivalent division problems (except when $a = b$), are students conversant with the notion of reciprocals? Are they able to see that 14 ÷ 2 yields the same result as 14 × ½ or 14 × 0.5? Assign *Practice: Multiplication and Division Patterns* (*Student Book*, p. 179).

Rationale

The *EMPower* authors recommend sticking to the meaning as essential grounding for students. If skipped or skimped on, students find themselves without a foundation for thinking through problems later. However, adults also need efficient ways to do the calculations. This lesson asks students to think about *why the procedures work* because that understanding will in turn highlight the properties of arithmetic, which are called upon in algebra. Interpreting the meaning of the problem, estimating with benchmarks, and attention to the math principles will shore up understanding and skill.

The goal is that every student end up with a reliable method he or she can use efficiently while understanding why the procedure works.

Get any group of adults in a room, ask them how they would do 8.214 ÷ 0.43, and you will be surprised at the variety of methods. Some result in a correct solution—others do not.

Here are some we have seen:

- Punch numbers into the calculator, starting with 8.214 (OK)
- Punch numbers into the calculator, starting with 0.43 (not OK)
- Set up the "house" division $0.43\overline{)8.214}$ (OK if using conventional U.S. method)
- Set up the "house" division $8.214\overline{)0.43}$ (not OK if using conventional U.S. method)
- Ask, "Can I do it the way I did in my country?" $0{,}43\overline{)8{,}214}$ (OK if using convention from France, Spain, and many other countries)
- Sigh and moan.

Then, since few people who work with numbers in their personal or working lives do long-hand division any more, hesitance and discomfort usually sets in as they start manipulating the numbers.

Traditional basic math workbooks for adults take a logical, step by step approach, starting with whole number divisors, stressing lining up points in the dividend and quotient. Then instruction moves on to moving points the same number of places. But what does decimal movement really mean? What principles of arithmetic permit that clever trick? **The essential idea is that we are rewriting the problem in another—equivalent—way.** And a way that is easier to work with.

$0.43\overline{)8.214}$ needs examination in light of the fact that any number times 1 is that number (multiplicative identity).

$$\frac{8.214}{0.43} = \frac{8.214}{0.43} \times 1 = \frac{8.214}{0.43} \times \frac{100}{100} = \frac{8214}{43}$$

↑

This choice of 100/100 is contrived to have the new divisor be a whole number, but any fraction equal to 1 could be used to get an equivalent problem.

Math Background

Prompts for division can include the division sign or the fraction bar or a situation that calls for splitting or dealing. If you do division on cue, you might not make time to make sense to the problem.

A numerate adult faced with a problem such as $8.214 \div 0.43$ takes a moment to think about what this "looks like" before jumping in. (Before you continue to read, do that yourself.)

One way to interpret this naked number problem is as "How many forty-three hundredths are in 8.214?" Then, we could make an estimate, relying on benchmarks. Perhaps, the answer has to be more than 16 because 0.43 is a little less than 0.50, or ½, and there are 16 halves in 8. Sometimes this "How many ___ in ___? interpretation is helpful. Other times, a sharing interpretation might be more apt.

There are other challenges embedded in this simple problem. Because division is not commutative, we have to be sure to distinguish between the divisor and the dividend. $8.214 \div 0.43$ does not have the same meaning as $0.43 \div 8.214$. Interpreting $0.43 \div 8.214$, perhaps one might see 0.43 cut into more than 8 parts (a sharing idea) or ask how many 8's in 0.43 (a small part of an 8). Ah yes, you can divide a big number into a smaller number! The answer to that question should be a small fractional amount. One way to picture it might be: "½ cut into about 8 parts is $\frac{1}{16}$, which is even less than 0.1." Now that we have a picture for the problem with the divisor and dividend reversed, we see that our answer will be very different.

Facilitation

Making the Lesson Easier

Reword problems to use smaller, friendlier numbers. Assign partners to complete practice pages. Post "How Many _____ in ____?" and "Split _____ into _____ groups" in the room as reminders.

Making the Lesson Harder

Assign the extension. For more challenge, ask students to draft a division problem from everyday life with decimals and to solve it using words, equations, and drawings to show how they know.

FACILITATING LESSON **11**

Applying Decimal Learning

Knowing when to use operations with decimals is the key.

Synopsis

Throughout this unit on decimals, students explored the operations. In this lesson, they will be asked to apply their knowledge of operations in various contexts to carry out a project. Each project is a real-life situation that calls on students to use decimals.

1. Student pairs choose a project to work on.
2. Students share their work and discuss how they used decimals and percents.
3. The class summarizes the lessons learned.

Objective

- Apply decimal operations and percents in real-life scenarios

Materials/Prep

- Full cereal box, if doing *Activity 2, Project 2* (*Student Book*, p. 193)
- Rulers (cm)
- Stock market performance data, if doing *Activity 2, Project 4* (*Student Book*, p. 196). Refer to instructions for using online resource **http://finance.yahoo.com/**

Opening Discussion

Set the stage for putting together all the concepts students have learned about decimals.

Today is a chance to put all the ideas from this unit together. Decimals and percents are often used when referring to data in exact measurements. These data are useful in making decisions. Today you will use decimals and percents to make decisions as well.

Then explain that a challenge in this lesson is figuring out which operations to use.

Keep track of how you decide what operation to use so we can talk about what you knew, how you tested a solution, and how you verified that it was the right way to go.

Remind them that it is as important to know when to use an operation as it is to know how to do the operation.

Activity 1: Number of the Day

Refer students to *Activity 1: Number of the Day* (*Student Book*, p. 190).

Today's number is 6.8. Write as many numerical expressions as you can that have the answer 6.8. Use all four operations. Use decimals and whole numbers. Write at least three equations with three decimals in them.

Students generate expressions. After a few minutes, ask how that went. Students may share some of their expressions. Then ask:

What do you have to keep in mind when you find the difference between two decimals? We can imagine a number line with the distance between the numbers being our target number, 6.8.

What addition problem is 10.7 – 3.9 related to? (3.9 + 6.8)

What multiplication problem is related to 34 ÷ 5? (5 × 6.8)

This kind of approach is an opportunity to support the more formal algebraic thinking students will need in future courses. If students are using "guess and check" approaches to finding expressions equal to 6.8 it's important to share this alternative and more sophisticated approach.

Activity 2: Pick a Project

Refer students to the four possible projects in *Activity 2: Pick a Project* (*Student Book*, pp. 191-197). Depending on time and students' interests, you may have them choose a project and pair with someone interested in the same project, or you may want the whole class to work through one or more of the projects.

Distribute materials. Review directions in the student materials. Briefly talk about each project. Be sure that students understand the tables shown for each project.

When students have completed their tasks, have each pair share their results.

Summary Discussion

Ask students to reflect on how they used decimals or percents to solve the problems in their projects.

What challenges did you face?

How did your understanding of decimals and percents come into play as you worked through the problems?

How did you decide what operation(s) to use?

What is something new that you learned?

Direct students to add any new terms they want to remember to *Vocabulary* (*Student Book*, p. 206) and to turn to *Reflections* (*Student Book*, p. 215) where students record what they want to remember.

Looking Closely

Observe whether students are able to

Apply decimal operations and percents in real-life scenarios

Look for evidence that students attach meaning to operating with decimals and percents. If students have trouble getting started, find out what aspects of the problems are similar to ones they have encountered in the past and ask how they have managed them. Ask them to spend some time talking with a partner: What do they want to find out? What do they expect to find out. What are some steps to take?

Rationale

Too often students emerge from math class and never use their hard-won skills. Instead, they use common sense or rely on other people or criteria and avoid making a decision based on numbers. This lesson and any extensions or projects with local relevance that can be brought to bear help forge the bridge between everyday life and school-based learning.

Context

Tailoring the projects to address local data-based decisions or construction projects may offer increasingly meaningful opportunities for problem solving with all the operations.

LESSON 11 IN ACTION

Here is how one teacher and his colleagues introduced Project 1: Applying for Life Insurance.

The teacher took the role of doctor and gave his colleague, playing the role of an insurance applicant, the results of her blood and urine tests. Using number lines on the wall, the two looked and found out which tests she "passed" and which she "failed."

The "doctor" gave out each of the blood test numbers on a sticky note and the insurance applicant had to plot them on huge number lines. The doctor said he wanted to make sure she understood why her life insurance premium would be so expensive unless she got herself in shape. As the doctor flirted with the patient, saying she was a "10.2" in his opinion, everyone in the room laughed.

New York City Teachers

Facilitating Closing the Unit: Put It Together

How well do you know your benchmarks?

Synopsis

In this lesson you give students a chance to show what they know about percents and the benchmark fraction one-tenth, as well as place value and operations with decimals. Students review their assignments and reflect on math problems they encounter in their everyday lives. They list two categories based on their past *Practice* pages and what they encounter daily: problems they feel confident solving and those that they still struggle to understand and solve. Then students complete the final assessment.

1. Students look through their work and list in two columns the problems they still struggle with and the problems they can now do with confidence.
2. Students take the assessment.

Objectives

- Identify problem areas students still need to work on
- Demonstrate ability to solve problems involving fractions, decimals, and percents
- Demonstrate conceptual understanding of operations involving decimals

Materials/Prep

- *Final Assessment, Appendices,* p. 137, one per student
- Students'assignments representing their work in this book
- *Final Assessment Checklist, Appendices,* p. 141, one per student to record your observations for individuals (optional)

Opening Discussion

During the time you've worked on this unit with students, they've seen fractions, decimals, and percents in many situations, both in class and likely outside of class too. Together, list some of the fractions, decimals, and percents they've noticed.

Ask for a few volunteers to share: What has changed for you in terms of noticing or using the fractions, decimals, and percents you encounter?

Activity 1: Self-Assessment

Ask students to turn to *Activity 1* (*Student Book*, p. 200) and to look back over their work in this unit.

On the left side, they should identify 3–5 problems with which they still have trouble.

On the right side of the page, they should identify types of problems they can now do with confidence and accuracy.

Students might simply write the page number and problem number for identification, or they might copy the problems from their assignments.

Activity 2: Final Assessment

Distribute a copy of the *Final Assessment* to every student. Review of the questions and instructions. Return to *Activity 2* (*Student Book*, p. 201) to reflect after the assessment.

Summary Discussion

Ask students to record their thoughts about the following.

What did you find out about your strong points?

What challenged you? What are you still curious about or frustrated by?

Direct students to *Reflections* (*Student Book*, pp. 215-216) to record final thoughts about what they learned from the project.

Collect the *Final Assessments*, and use the *Assessment Checklist* to give students a sense of how well they did and where their skills are strong and where they need development.

Appendices

Name ______________________________ Date __________

Split It Up: More Fractions, Decimals, and Percents

INITIAL ASSESSMENT

Task 1: Number of the Day

Write as many numerical expressions as you can that have the answer 10.5.

Task 2: Tips

Nan waits tables at a steak house. At the restaurant, tips are divided up. Nan keeps a percent. Below is the breakdown.

Calculate everyone's share of the tips for Tuesday night.

TUESDAY NIGHT TOTAL TIPS: $240	
	How much did each person make?
1. 5% to the hostess	
2. 5% to the bartender	
3. 10% to the busboys	
4. 25% to the captain	
5. 25% to the back waiter	
6. 30% to the front waiter (Nan)	

Task 3: Frames

Two mirrors are pictured below. How much framing is needed to go around each one? The sides of each mirror are equal.

1. 

Side = 11.25"

Total needed: ___________

2. 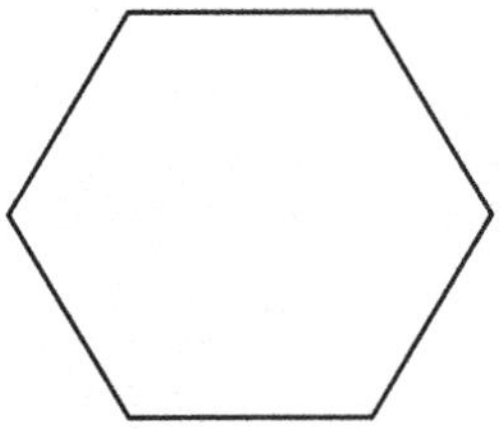

Side = 12.9 cm

Total needed: ___________

Task 4: Write it in Numbers

Write a whole number or decimal answer.

1. 5 hundredths ______________________
2. quarter of a million ______________________
3. 0.5 million ______________________
4. 75K ______________________
5. $7(1) + 6\left(\frac{1}{10}\right) + 2\left(\frac{1}{100}\right)$ ______________________
6. seven and fifty-one thousandths ______________________

7. seventy and fifty-one hundredths ____________________

8. seven million, fifty-one thousand ____________________

Task 5: More Than, Less Than, or the Same As

Insert the correct words in the blanks.

1. 0.4 is ____________________ 0.22

2. 4 hours 25 minutes is ____________________ 4.25 hours

3. $\frac{1}{10}$ is ____________________ 10^{-1}

4. 0.091 is ____________________ 0.87

Task 6: Fill in the Equivalents

Use what directions you know to fill in the blanks in the table.

FRACTION NAME IN NUMBERS OR WORDS	PERCENT EQUIVALENT	DECIMAL EQUIVALENT
1. One-half		
2. Two-fourths		
3. Four-fourths		
4. One-tenth		
5.	12.5%	
6.		0.005
7. Four-fifths		

Task 7: Zinc Coating

Every penny made gets about 0.13 mm of zinc coating. The numbers in a–f are measures of the coating on six pennies.

a. 0.1 mm **b.** 0.11 mm **c.** 0.2 mm **d.** 0.25 mm **e.** 0.8 mm **f.** 0.08 mm

Place the numbers on the number line. (This is a greatly enlarged image of 1 millimeter!)

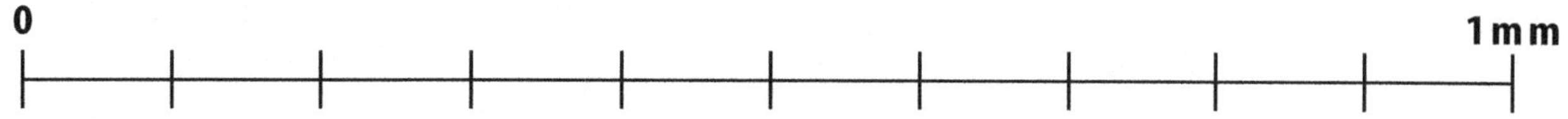

Task 8: Tolerances

Fill in the blanks to make the sentences below true. Use the measurements in Task 7 (a–f) to answer the questions about readings that are within and outside the tolerance of ± 0.04 mm.

1. The penny with the coating closest to 0.13 mm got a coating of ____mm.
2. The coating ranged from a thin coat of ______ to a thick coat of ______.
3. A coating of ______ is 10 times thicker than a coating of ______.
4. A coating of ____ is nearly twice what it should have been.
5. The following measurements were within the tolerance of ± 0.04 mm: ____________________

Task 9: Place the Point

Without calculating, place the decimal point where it belongs in the answer.

1. 0.96 + 0.2 = 1 1 6
2. 9.6 – 0.02 = 9 5 8
3. 0.96 × 2 = 1 9 2
4. 0.096 ÷ 8 = 0 0 1 2
5. 1266 – 3.02 = 1 2 6 2 9 8
6. 1.208 + .30 = 1 5 0 8

Task 10: S&P Drop

The S&P 500 stock index is at 1500 points.

1. If the value were to decrease by $\frac{1}{2}$ of 1% tomorrow, by how many points would it fall?

2. How did you figure it out?

Task 11: A Smaller Box

A paperclip box measures 7 centimeters (cm) in length, 4.5 cm in width, and 1.75 cm in height. If the box were shrunk by 10% in each dimension, it would measure:

1. ______ cm in length

2. ______ cm in width

3. ______ cm in height

Task 12: Human Hair

An average strand of human hair is between 0.003 and 0.007 of an inch wide. About how many strands of hair 0.005 inches in width would you need to line up side by side to make an inch?

How did you figure that out?

Task 13: Card Thickness

Most plastic ID cards in the U.S. are 3 hundredths of an inch thick. How many ID cards would you need to stack together to get a pile that is 10 inches high?

How did you figure that out?

INITIAL ASSESSMENT CHECKLIST

Use a ✓, ✓+, or ✓– to assess how well students met each skill. Note strengths as well as areas for improvement.

Student's Name____

Item/Task	Skills	Lesson Taught
1. Number of the Day	Writing expressions or equations with decimal and whole numbers	All
2. Tips	Finding percent amounts	*3, 4*
3. Frames	Adding or multiplying with decimal or fraction amounts	*8, 9*
4. Write It in Numbers	Place value	All
5. More Than, Less Than, or the Same As	Comparing decimal and fraction quantities	All
6. Fill in the Equivalents	Naming fraction, decimal, and percent equivalents	*1, 2*
7. Zinc Coating	Ordering and comparing decimal amounts; interpreting a line with unmarked increments	*1-7*
8. Tolerances	Comparison of decimal amounts; addition and subtraction with decimals	*2, 6*
9. Place the Decimal Point	Place value	*8, 9, 10*
10. S&P Drop	Calculating percents; explaining a stragegy for calculating a drop expressed in percents	*3, 4*
11. A Smaller Box	Operations with decimal numbers	*8*
12. Human Hair	Calculating with numbers in the thousandths	*7, 8, 9*
13. Card Thickness	Calculating with numbers in the hundredths	*7, 8, 9*

Strengths and Areas for Improvement

Name ______________________ Date __________

Split It Up: More Fractions, Decimals, and Percents

FINAL ASSESSMENT

Task 1: Number of the Day

Write as many numerical expressions as you can that have the answer 10.5.

Check your list: Did you use all four operations (+, ×, –, ÷), fractions, decimals, percents, mixed numbers, and exponents?

Task 2: Vegetable Sale

A group of friends agreed to raise and sell vegetables. They arranged to share the profits.

Calculate everyone's share of the profit collected August 31st.

TOTAL SOLD—MARKET DAY AUG 31: $250 PROFIT	
PERCENT/TASK	HOW MUCH DID EACH PERSON MAKE?
1. 5% Ordered seeds (Nuku)	
2. 10% Land owner (Kate)	
3. 20% Harvested, prepped for sale (Ben)	
4. 25% Staffed market table (Kira)	
5. 40% Planted, watered, weeded (Petra)	

Task 3: Frames

Two stained glass panels are below. How much framing is needed to go around each one? The sides of each panel are equal.

1. 

Side = 6.25″

Total needed: ___________

2. 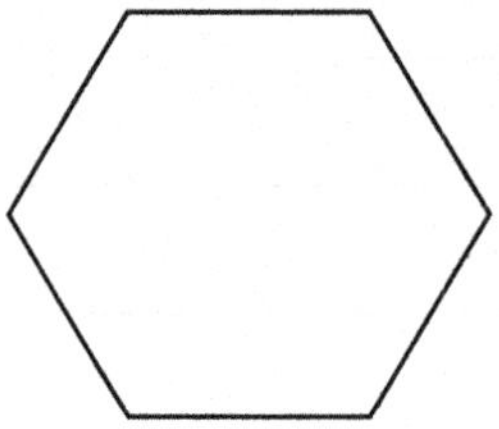

Side = $5\frac{3}{8}$″

Total needed: ___________

Task 4: Write It in Numbers

Use whole numbers and decimals.

1. 3 hundredths ______________________________
2. Three quarters of a million ______________________________
3. 0.25 million ______________________________
4. 75K ______________________________
5. 9(1) + 5 (1/10) + 2 (1/100) ______________________________
6. four and twenty-one thousandths ______________________________
7. thirty and twenty-one hundredths ______________________________
8. two million, twenty-one thousand ______________________________

Task 5: More Than, Less Than, or the Same As

Insert the correct words.

1. 30% is ____________________ $\frac{1}{3}$

2. 75 minutes is ____________________ .75 hours

3. $\frac{1}{10}$ is ____________________ 10^{-1}

4. 0.075 mg is ____________________ an aspirin dose of .325 mg

Task 6: Mole Removal

Nina needed a mole removed. She heard many different opinions on how deep the cut could be. Use the number line to show the different measurements Nina was told.

a. 0.91 mm **b.** 0.16 mm **c.** 9 mm **d.** 4.5 mm **e.** 0.09 mm **g.** 0.08 mm

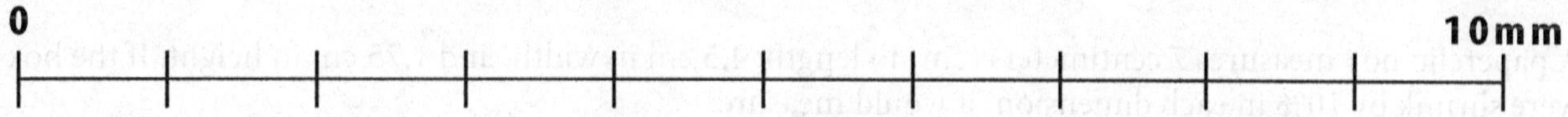

Task 7: Surgical Cuts

Fill in the blanks to make the sentences below true. Use the numbers from Task 6.

1. The shallowest case is a superficial cut of ______; the deepest cut is ______.

2. ______ is about 10 times deeper than ______.

3. ______ is half as deep as ______.

Task 8: Place the Decimal Points

Without calculating, place the decimal point where it belongs in the answer.

1. $0.96 + 0.2 = 1\ 1\ 6$

2. $9.6 - 0.02 = 9\ 5\ 8$

3. $0.96 \times 2 = 1\ 9\ 2$

4. $0.096 \div 8 = 0\ 0\ 1\ 2$

Task 9: S&P Increase

The S&P 500 stock index is now at 1500 points.

1. If the value goes up 1/2%, what is the amount of the increase?

2. How did you figure it out?

Task 10: A Smaller Box

A paperclip box measures 7 centimeters (cm) in length, 4.5 cm in width, and 1.75 cm in height. If the box were shrunk by 10% in each dimension, it would measure:

1. _____ cm in length
2. _____ cm in width
3. _____ cm in height

Task 11: Aluminum Cans

In the 1970's, aluminum cans were about as thick as aluminum gutters: 0.015 inches. How many layers at that thickness would it take to make an inch?

Task 12: Comparing Thickness

In 2013, the thickness of a soda can was 0.0047". Describe the change in thickness compared to a 1970's can that was 0.015".

FINAL ASSESSMENT CHECKLIST

Use a ✓, ✓+, or ✓– to assess how well students met each skill. Note strengths as well as areas for improvement.

Student's Name____

Item	Skills	Lesson Taught
1. Number of the Day	Writing expressions or equations with decimal and whole numbers	All
2. Vegetable Sale	Finding percent amounts	*3, 4*
3. Frames	Adding or multiplying with decimal or fraction amounts	*8, 9*
4.Write It in Numbers	Place value	All
5. More Than, Less Than, or the Same As	Comparing decimal and fraction quantities	All
6. Mole Removal	Interpreting a line with unmarked increments; comparing decimal amounts	*1-7*
7. Surgical Cuts	Comparison of decimal amounts	*2, 6*
8. Place the Decimal Point	Place value	*8, 9, 10*
9. S&P Increase	Calculating percents; explaining a strategy for calculating a drop expressed in percents	*3, 4*
10. A Smaller Box	Operations with decimal numbers	*8*
11. Aluminum Cans	Calculating with numbers in the thousandths	*7, 8, 9*
12. Comparing Thickness	Comparing numbers in the thousandths	*7, 8, 9*

Strengths and Areas for Improvement

____

____

____

____

____

____

Blackline Master 1: Stamps

Blackline Master 2: Match Cards

0.1		
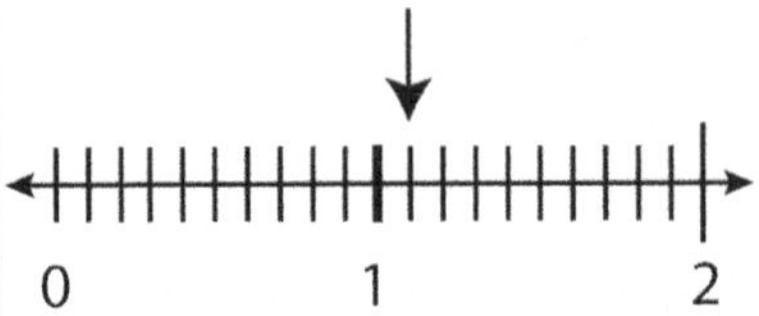		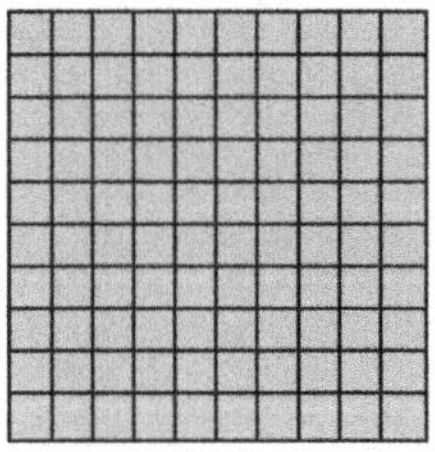
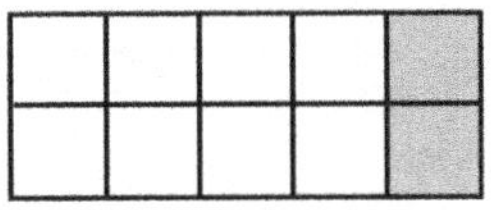	$\frac{1}{10}$	1.0
$\frac{10}{1}$	10%	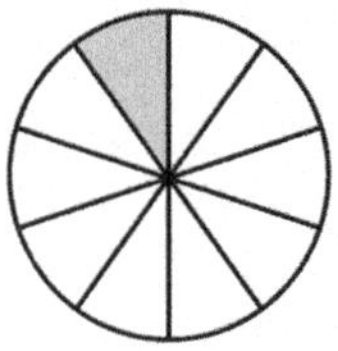
$\frac{2}{10}$	.2	100%
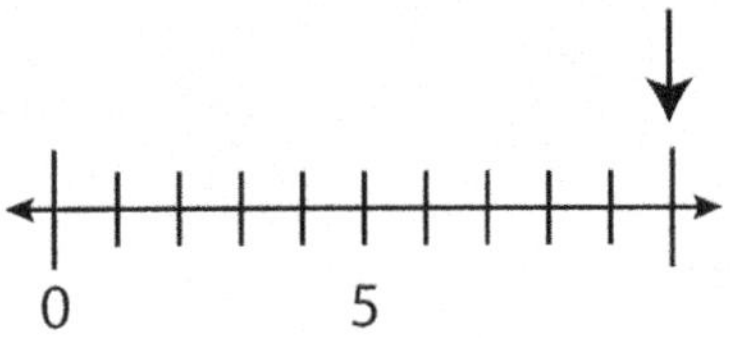	one-tenth	point one
1.1	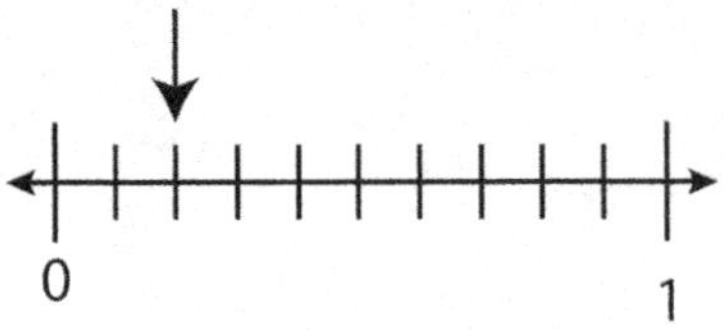	10.0

Blackline Master 3: Two Out of Twenty

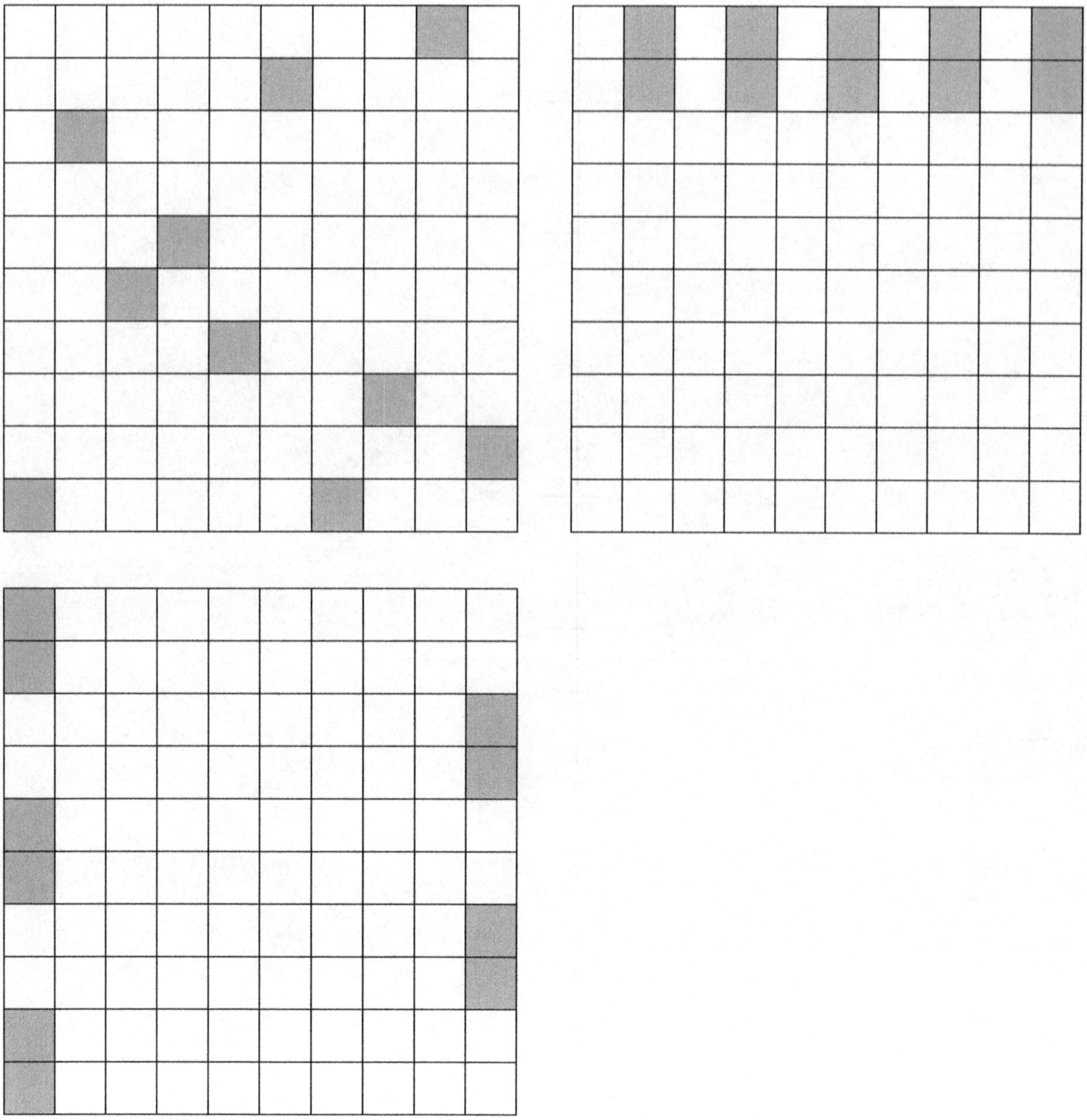

Blackline Master 4: One Percent on the Number Line

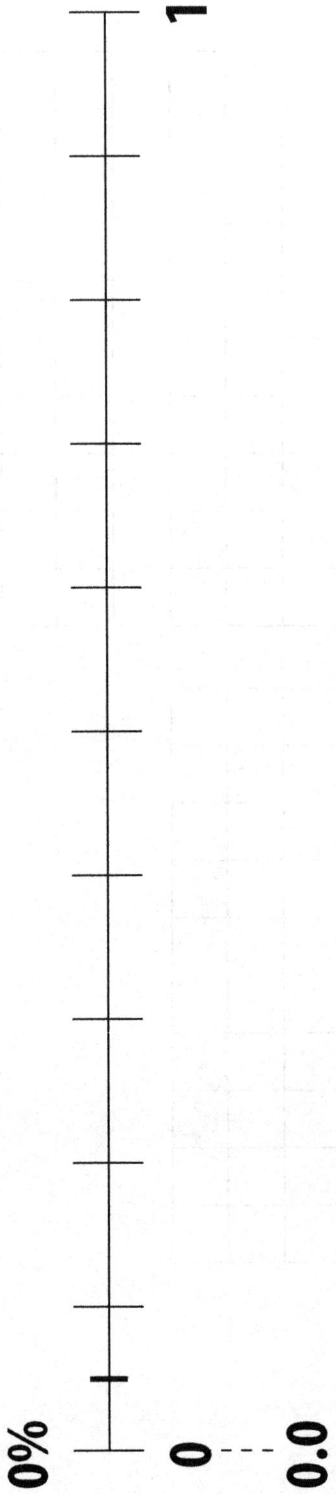

Blackline Master 5: Money Strips

$1	$1	$1	$1	$1
$1	$1	$1	$1	$1
$1	$1	$1	$1	$1
$1	$1	$1	$1	$1
$1	$1	$1	$1	$1
$1	$1	$1	$1	$1
$1	$1	$1	$1	$1
$1	$1	$1	$1	$1
$1	$1	$1	$1	$1
$1	$1	$1	$1	$1

Blackline Master 6: Ten Cards

$\frac{4}{5}\%$	$\frac{4}{5}$
0.08	80%
$\frac{8}{10}$	$\frac{8}{100}$
0.8%	$\frac{80}{100}$
0.80	4.5%

Blackline Master 7: Thousandths Cards, A–H

A

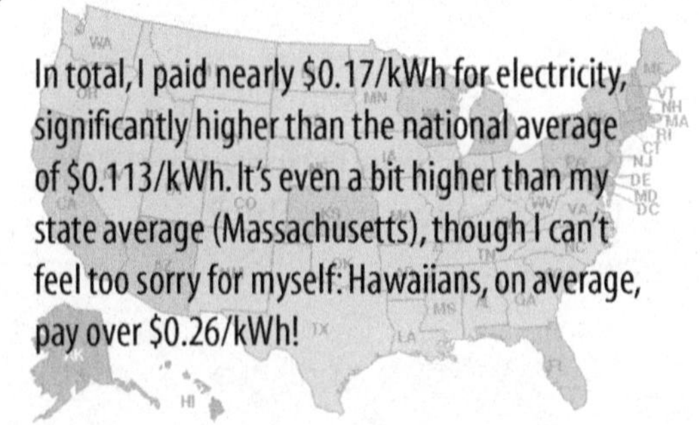

In total, I paid nearly \$0.17/kWh for electricity, significantly higher than the national average of \$0.113/kWh. It's even a bit higher than my state average (Massachusetts), though I can't feel too sorry for myself: Hawaiians, on average, pay over \$0.26/kWh!

B

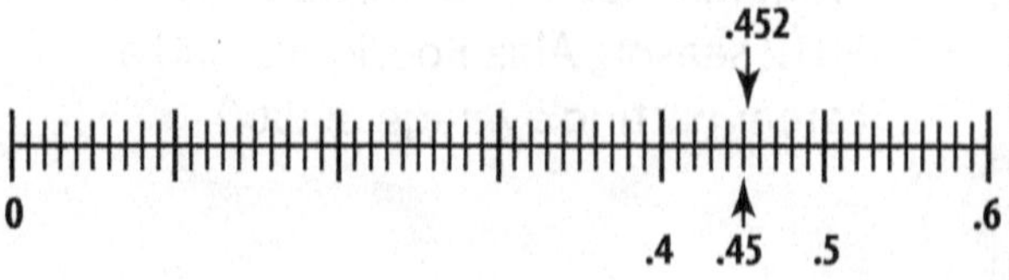

C

The number π (pi) is a mathematical constant that is the ratio of a circle's circumference to its diameter, and is approximately equal to 3.14159.

D

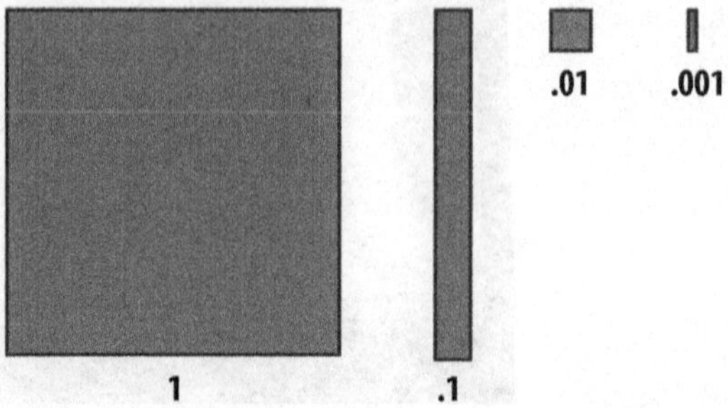

E

F

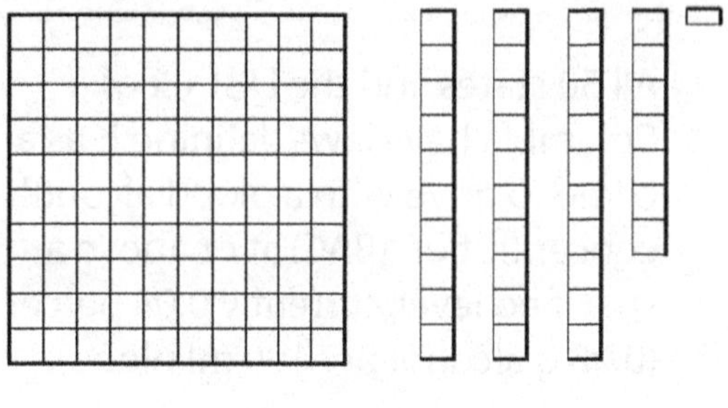

1.375

G

$\frac{1}{3} = 0.3333333 \ldots$

(Try this on a calculator)

H

$\frac{5}{8} = 0.625 = 62.5\%$

Blackline Master 7: Thousandths Cards, I–N

I

From his rookie year through to the 2012 season, Alex Rodriguez had a career batting average of .300.

J

A strand of hair is between 0.003 to 0.007 of an inch thick.

K

L

All 50 states and the District of Columbia have laws defining it as a crime to drive with a blood alcohol concentration (BAC) at or above a specified level, currently 0.08 percent (0.08 g alcohol per 100 ml blood).

M

The U.S. economy is expected to grow by 2.5 percent in 2013, improving to 3.5 percent growth in 2014, top Fed official Charles Evans said on Monday.

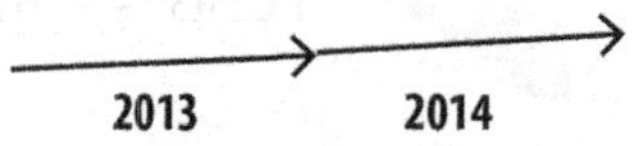

N

	December 2011			December 2012		
	U.S.	Chicago area	Percent difference	U.S.	Chicago area	Percent difference
Electricity per kWh	$0.127	$0.147	15.7	$0.127	$0.152	19.7
Gas per gallon	3.329	3.432	3.1	3.386	3.541	4.6
Utility (piped) gas per therm	1.034	0.822	-20.5	1.004	0.788	-21.5

Note: A positive percent difference measures how much the price in the Chicago area is above the national price, while a negative difference reflects a lower price in the Chicago area.

Blackline Master 8: 0–9 Cards

1	2	3
4	5	6
7	8	9
	0	

Blackline Master 9: 100-Block Grid

Blackline Master 10: 100-Block Grids

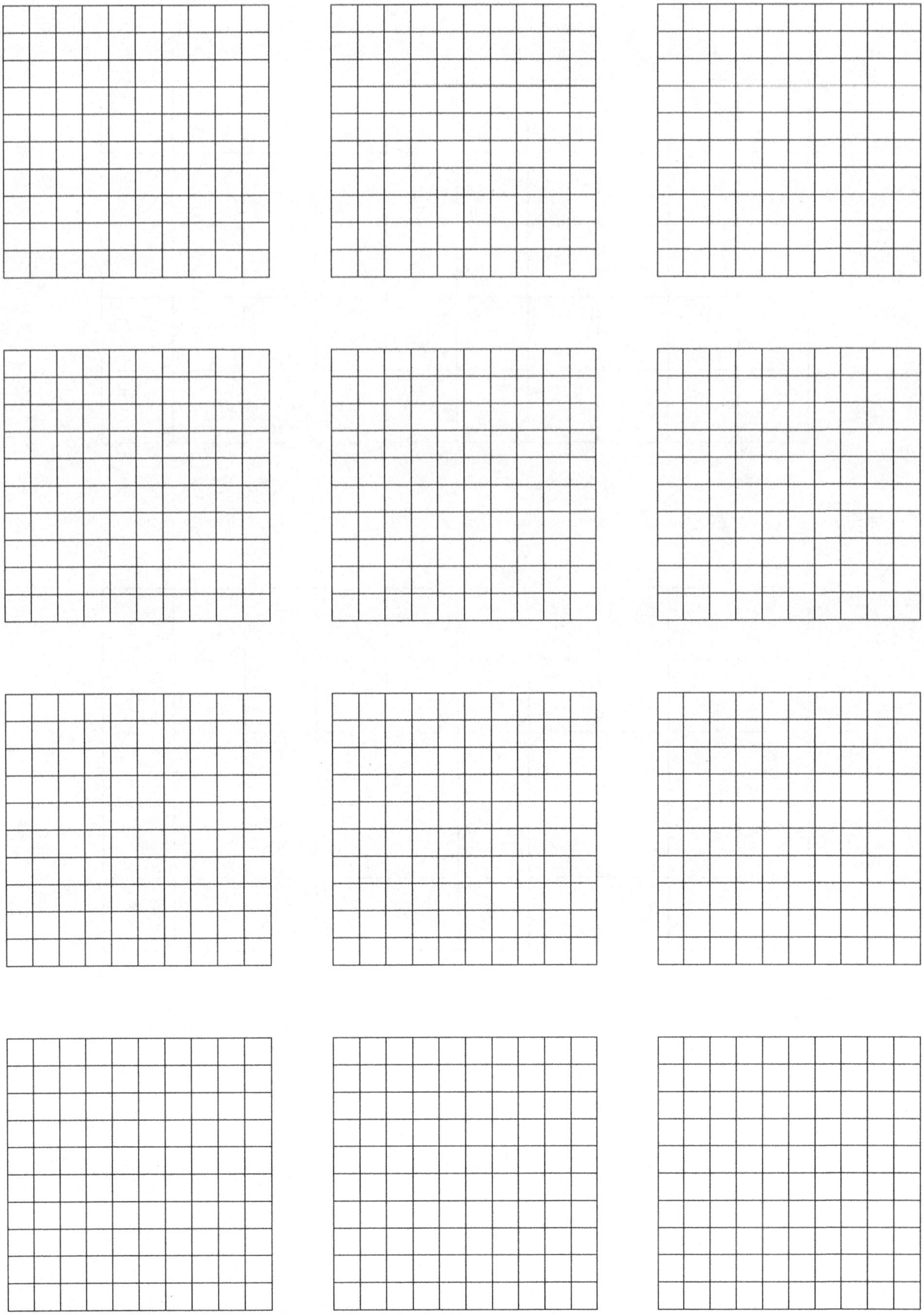

Blackline Master 11: 50-Block Grids

Glossary

Reminder: Students generate their own definitions for terms as they arise in class, using language that makes sense to them. However, to help you guide the discussion, we include mathematical definitions for most of the terms.

Italicized terms are *not* included in the *Student Book* but are included in the *Teacher Book* as background information. The lesson number in which the term first appears is in parentheses following the term.

area model (9)—a way to show multiplication as square units of surface area on a grid.

benchmark (1)—a familiar number used as a point of reference. For example, ½ is a benchmark fraction because it can be used easily as a point of reference for comparing quantities greater than or less than a half.

decimal (Opening)—a number with one or more digits to the right of the decimal point. For example, 5.83 and 0.54 are decimal numbers.

decimal point (2)—a dot or point used to separate the whole and fractional portions of a number.

decompose (5)—to take apart numbers, usually separating by place value, ones and tens, for example.

denominator (1)—the part of a fraction written below the line that indicates the number of parts into which one whole is divided.

digits (2)— single whole number, 0–9, in a number 10 or larger. For example, 15 has two digits.

estimate (Opening)—to calculate approximately.

expanded notation (7)—a way to write numbers that shows the place value of each digit; for example, $4{,}265 = 4 \times 1{,}000 + 2 \times 100 + 6 \times 10 + 5 \times 1$

exponents (9)—a way to indicate a number times itself, notated with a superscript, as in 10^2.

fraction (Opening)—a number that names a part of a whole or a part of a group.

gauge (2)—instrument for measuring. Gauges sense thickness, weight, temperature and have different levels of sensitivity, so they may be labeled in increments of whole numbers or tenths or smaller units. Gauges may be analog or digital.

gross (4)—the amount of money acquired before any costs are deducted.

hundredths (6)—the 100 equal parts of any whole, written to the first and second decimal place, equivalent to percentages.

median (7)—one way to find an average. After ordering a list of values, in an odd set of numbers, the median is the middle value. In an even set of numbers, it is the (mean) average between the two numbers closest to the middle of the list.

multiples of 1% (5)—1%, 2%, 3%, 4%, etc.

multiples of 10% (3)—20%, 30%, 40%, 50%, 60%, 70%, 80%, 90%, 100%, etc.

net (5)—the amount of money kept after costs are deducted. In *Lesson 5*, net is equivalent to take-home pay.

numerator (3)—the part of a fraction that is above the line and signifies the number to be divided by the denominator.

one percent, 1% (4)—one out of 100.

one-tenth, 1⁄10 (1)—one out of ten objects or collections of objects. For example, if 1⁄10 of the people in a group of ten are vegetarians, it means that one is a vegetarian and the other nine are not.

one-hundredth, 1⁄100 (6)—one of 100 equal parts that make a whole. For example, 2 is 1⁄100 of 200, because 1 out of 100 is 1⁄100, 2⁄200 is the same part-whole ratio.

operation sense (11)— understanding the meanings and models of operations, the real world situations they connect with, and the symbols that represent them.

part (Opening)—a portion related to the whole. For example, in ¾, "3" is the part and "4" is the whole.

percent (**Opening**)—the ratio of a number to 100; percent means "per hundred." For example, 25% is 25 out of 100, or 25 per 100.

product (9)—the name for the answer that results from multiplication.

rectangular array (9)— an arrangement of objects into rows and columns that form a rectangle. For example, a 24-pack of water, arranged in 6 rows by 4 bottles in each row.

rounding (**3**)—a type of estimation or approximation that makes numbers easier to use. For example, if you are counting paper clips and there are 148, you might *round* to 150, the nearest ten. With decimals, if you are considering 1.77, you might *round* to 2.

square (9)—a number times itself, any number written with the exponent of 2, as in x^2.

ten percent, 10% (**2**)—10 out of 100.

thousandths (7)— one of a thousand equal parts.

tithe (**1**)—to pay or contribute one-tenth of some amount (e.g., income).

whole (**Opening**)—the entire object, collection of objects, or quantity being considered; 100%.

Answer Key

Opening the Unit

Activity 1: Newspaper and Magazine Search

Answers will vary.

Activity 2: Initial Assessment

Task 1: Number of the Day

Answers will vary.

Task 2: Tips

	Tuesday night tips
1. 5% to hostess	$12
2. 5% to bartender	$12
3. 10% to busboys	$24
4. 25% to captain	$60
5. 25% to the back waiter	$60
6. 30% to the front waiter (Nan)	$72

Task 3: Frames

1. $11.25 \times 4 = (11 \times 4) + (0.25 \times 4) = 44 + 1 = 45''$
2. $12.9 \times 6 = (12 \times 6) + (0.9 \times 6) = 72 + 5.4 =$ 77.4 cm

Task 4: Write It in Numbers

1. 0.05
2. 250,000
3. 500,000
4. 75,000
5. 7.62
6. 7.051
7. 70.51
8. 7,051,000

Task 5: More Than, Less Than, Or the Same As

1. more than
2. more than
3. the same as
4. less than

Task 6: Fill in the Equivalents

Fraction name in numbers or words	Percent equivalent	Decimal equivalent
1. One-half	50%	0.5 or .5
2. Two-fourths	50%	0.5 or .5 or 0.50
3. Four-fourths	100%	1 or 1.0
4. One-tenth	10%	0.1 or .1 or .10
5. One-eighth or 1/8 or 125/1,000	12.5%	0.125 or .125
6. Five-thousandths or 5/1,000 or 1/200	.5%	0.005
7. Four-fifths	80%	0.8 or .8 or .80

Task 7: Zinc Coating

On the number line, the numbers should be approximately placed and ordered (from left to right): 0.08. 0.1, 0.11, 0.2, 0.25, 0.8.

Task 8: Tolerances

1. 0.11
2. 0.08, 0.8
3. 0.8, 0.08
4. 0.25
5. 0.1, 0.11

Task 9: Place the Point

1. 1.16
2. 9.58
3. 1.92
4. 0.012

5. 1,262.98
6. 1.508

Task 10: S&P Drop

1. 7.5 points
2. Answers will vary, but one possible way is:

   $\frac{1}{2}$ of 1% of 1500

   10% of 1500 = 150

   1% of 1500 = 15

   $\frac{1}{2}$ of 1% of 1500 = $\frac{1}{2}$ of 15 = 7.5

Task 11: A Smaller Box

1. 6.3 cm in length
2. 4.05 cm in width
3. 1.575 cm in height

Task 12: Human Hair

You would need 200 strands of hair. Possible explanation: 0.01 would be 100 hairs. I want to know about a space half that size, 0.005. I need to double 100. I would need 200 strands of hair.

Task: 13 Card Thickness

0.01 would be 100 in an inch. Possible explanation: Three times that, 0.03, means 1/3 of 100 in an inch, or 33.3. I need 10 inches. 33.3 × 10 is 333.3. About 333 cards. (Also acceptable: 333.33, or 333.3 or 333 1/3.)

Lesson 1: One-Tenth

Activity 1: Building on the Fractions Strips

Part 1: Labeling Fraction Strips

1. a. 2/10 = 1/5 ; 4/10 = 2/5; 5/10 = 1/2 = 2/4 = 3/6 = 4/8 = 6/12; 6/10 = 3/5; 8/10 = 4/5
   b. Possible answers: The fifths numerator doubled in the numerator for tenths; for 1/2, the pattern is 1, 2, 3, 4, 5 for the numerators and the denominators double each time 2, 4, 6, 8, 10.
2. a. 1/10, 1/8, 1/6, 1/5, 1/4, 1/3, 1/2
   b. For fractions with a numerator of 1, the larger the denominator, the smaller the fraction.
3. a. 1/8; more than 1/10
   b. 1/3; more than 3/10 (2/6, 4/12, and 5/16 are also close to 3/10 but are not benchmark fractions)
   c. 3/4 or 2/3; 3/4 is more than 7/10, 2/3 is less than 7/10
   d. 3/10

Part 2: Labeling Fraction Strips with Decimals

1. The halves should be labeled 0.5 and 1.0. The fourths fraction strip should be labeled 0.25, 0.50, 0.75, 1.00. The fifths should be labeled 0.2, 0.4, 0.6, 0.8, and 1.0. The tenths fraction strip should be labeled with tenths, left to right, from 0.1 to 1.0.
2. a. True
   b. True
   c. True
   d. False. Possible reasoning: 0.4 represents four-tenths. When I hold 4/10 up to 1/4, I see that it is larger and cannot be equivalent.
   e. False. Possible reasoning: 0.2 represents two-tenths. When I hold 2/10 up to 1/2, I see that 1/2 is much larger.
   f. False. Possible reasoning: 0.5 represents five-tenths which is equivalent to 1/2. When I compare 1/2 to 1/5, I see that 1/2 is larger and cannot be equivalent.

Activity 2: Show Me 1/10 Stations

See *Looking Closely* for suggestions on what to look for as students describe their data in fractional terms.

Station 1:

1. 1/10 of 100 is 10.

Station 4:

1. 2 ½ cents or $0.025.
2. The pennies couldn't be split into 10 equal groups.

Station 5:

1. 40 stamps are in the entire collection.

Practice: I Will Show You 1/10!

1. One dime represents one-tenth.
2. 1 year
3. 2 pages
4. 4 beads
5. 450 voters

Practice: Containers

1. Possible explanation: I saw that there are ten rows of ten squares. One row of 10 is one of the ten rows, or one-tenth. (1/10)

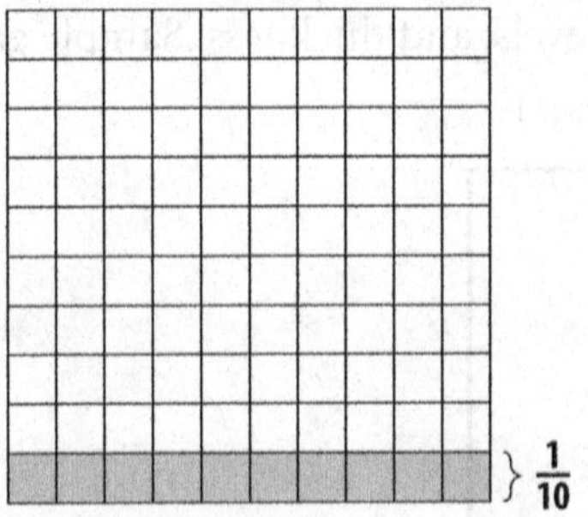

2. Possible explanation: I marked the container with 9 lines spaced evenly apart to create 10 sections. I filled the bottom one to show one out of 10, or 1/10.

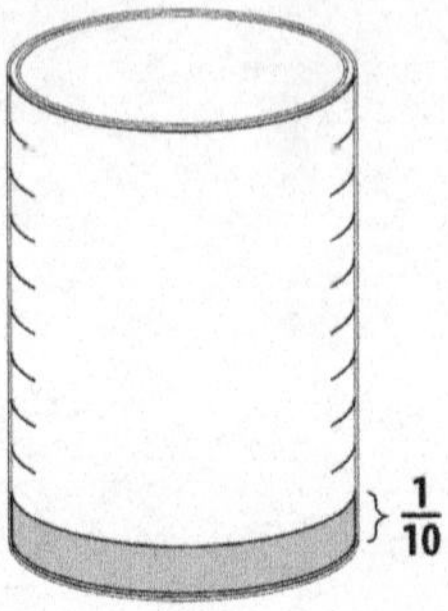

3. Possible explanation: I imagined making a loaf of bread and cutting it into 10 pieces. I shaded one out of 10 of the pieces to show one-tenth (1/10).

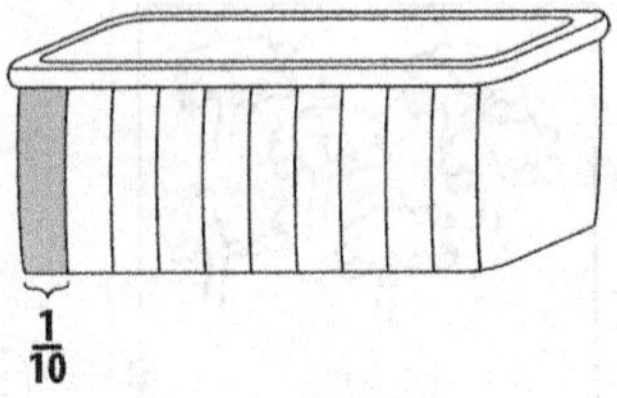

4. Possible explanation: Between the fill line and the bottom of the jar, I marked 9 lines to creat 10 sections. I filled the bottom of the jug to show one out of ten.

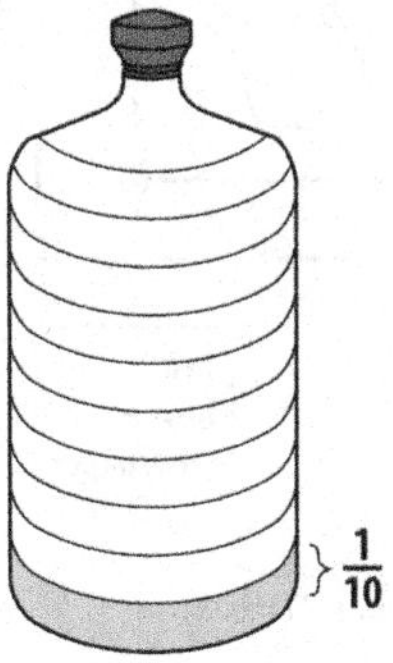

Practice: More or Less?

Explanations may vary.

1. More than one-tenth; 10/10 is the whole usual walk. 1/10 would be a tenth of the usual walk. Yesterday she walked 4/10 which is 3 more than 1/10.
2. Less than one-tenth; 230 is the total number of pages. 230 divided by 10 is 23, so 23 is 1/10 of 230. If she read 12 pages, she read less than one-tenth.
3. More than one-tenth; Half a mile is five-tenths and five-tenths is larger than one-tenth.
4. More than one-tenth.

Extension: More Tenths

Explanations may vary.

1. $180. One-tenth is $60, so 5/10, or 1/2, is $300. She has saved $120 already, so she needs $180 more.
2. If they sell 240 cups in 1/10 of a day, then they sell 2,400 cups during the whole day. One-quarter of 2,400 cups is 600 cups.

Test Practice

1. (c)
2. (c)
3. (c)
4. (b)
5. (d)
6. $1,450

Lesson 2: More About One-Tenth

Activity 1: One-Tenth Match

See *Looking Closely* for suggestions on what to look for as students determine what fraction their cards represent.

Activity 2: Hard Questions

Answers will vary. See teacher notes for suggestions for Homework Problems 1-3.

Activity 3: Why Is 10% Equal to 1/10?

Answers will vary.

Math Inspection: Entering Numbers into Machines

1.

Calculator display	Dollars and cents
a. 0	0 dollars and 0 cents
b. 7 or 7.	7 dollars and 0 cents
c. 7 or 7.	7 dollars and 0 cents
d. 7 or 7.	7 dollars and 0 cents
e. 7 or 7.	7 dollars and 0 cents
f. 7 or 7.	7 dollars and 0 cents
g. 0.7	0 dollars and 70 cents
h. 0.7	0 dollars and 70 cents
i. 0.7	0 dollars and 70 cents
j. 0.07	0 dollars and 7 cents
k. 0.07	0 dollars and 7 cents
l. 0.7	0 dollars and 70 cents
m. 7.7	7 dollars and 70 cents
n. 7.07	7 dollars and 7 cents
o. 7 or 7.	7 dollars and 0 cents
p. 70.07	70 dollars and 7 cents
q. 70 or 70.	70 dollars and 0 cents
r. 70 or 70.	70 dollars and 0 cents
s. 0.7	0 dollars and 70 cents

2. Students should notice that trailing zeroes disappear once you hit enter and zeroes in front of whole numbers do not appear on the calculator. They may notice that in problems like example (i), with values less than one, the calculator displays a 0 in the ones place. Depending on the calculator type, the decimal point may not appear until the person using it enters a decimal point.
3. Students will verify with the chart.

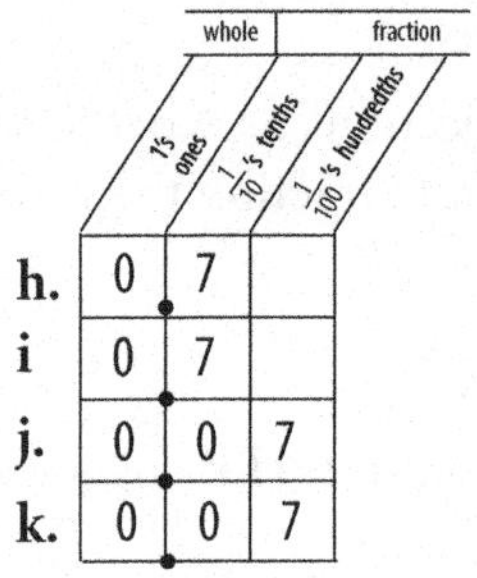

4. Possible answer: The calculator displays only zeroes that are needed to know the place value of the digits.
5. The calculator eliminates trailing zeroes or zeroes in front of whole numbers.

Practice: Make a Gauge

1. Students must choose a decimal larger than zero and less than 5.
2. Answers will vary. Gauges measure pressure, volume, levels, and thickness. Sample gauge:

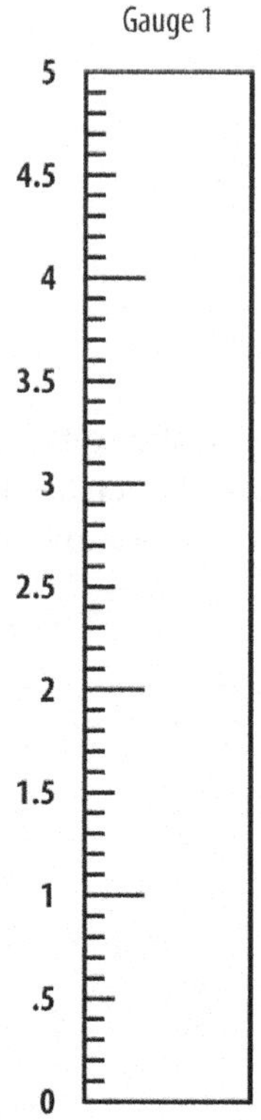

Practice: Location, Location, Location

1.

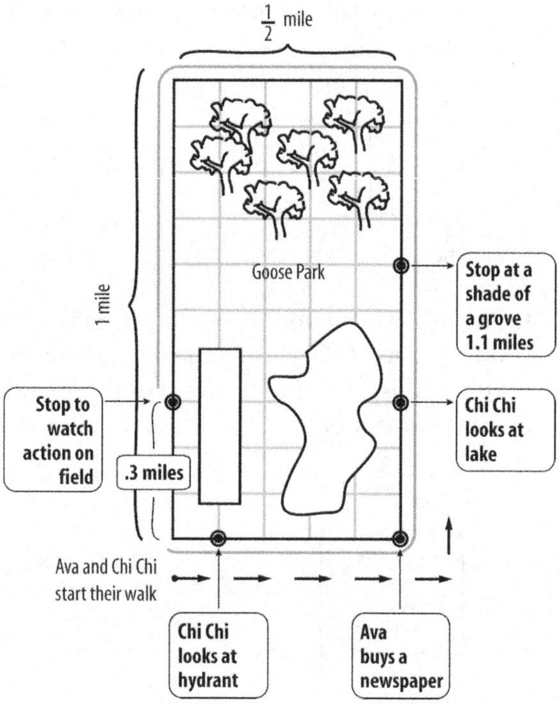

d. They have walked 1.1 miles.

f. They need to walk 0.3 miles to get back to their starting point.

Practice: Gaining Weight

	original weight	0.1 of original	increased by 0.1	decreased by 0.1
Claudia	155	15.5 lb.	170.5 lb.	139.5 lb.
Salvatore	200	20 lb.	220 lb.	180 lb.
Jackie	110	11 lb.	121 lb.	99 lb.
Sue	180	18 lb.	198 lb.	162 lb.
Calvin	250	25 lb.	275 lb.	225 lb.
Lamika	135	13.5 lb.	148.5 lb.	121.5 lb.
Igor	140	14 lb.	154 lb.	126 lb.

Practice: Digits, Places, and Points

1. a. millions, hundred thousands, ten thousands, one thousands, hundreds

2. a.

whole							fraction		
1,000,000's millions	100,000's hundred thousands	10,000's ten thousands	1,000's one thousands	100's hundreds	10's tens	1's ones	$\frac{1}{10}$'s tenths	$\frac{1}{100}$'s hundredths	$\frac{1}{1,000}$'s thousandths
					9	7			
			7	8	0	5			
	7	5	0	0	0	0			

b. ones, thousands, hundred thousands

c. thousands, hundred thousands

3. a. The point separates the wholes from the parts of a whole that are smaller than one.

b

whole							fraction		
1,000,000's millions	100,000's hundred thousands	10,000's ten thousands	1,000's one thousands	100's hundreds	10's tens	1's ones	$\frac{1}{10}$'s tenths	$\frac{1}{100}$'s hundredths	$\frac{1}{1,000}$'s thousandths
						9	7		
				7	8	0	5		
		7	5	0	0	0	0		

c. tenths, hundreds, ten thousands

4. a. They have the same value. They both represent 9/10; they are in the same place on the place-value chart.

b. They have the same value. Possible answer: each 9 is in the ones place and in both cases, there are no tenths.

c. No. Possible answer: 0.9 lies to the right of the decimal point and is in the tenths place, and there is no whole part. 1 is to the left of the decimal point and is in the ones place, so because it has a whole unit, it must be bigger.

5. .1, .8, 1., 1.3, 2, 3.0, 3.4, 12.5, 13, 125

6. 98.8, 98.9, 99.0, 99.1

7. a. $1,000,000; $1,000,000,000; $1,000,000,000,000.

b. Each label (like millions and billions) adds on three more zeroes. When you expand to the next label, you have 1,000 times more: a million is 1,000 times 1,000; a billion is 1,000 times a million.

8. a. thousand;

b. thousand;

c. five hundred;

d. $500,000;

e. thousand;

f. one thousand;

g. $500,000,000

9 a. Powerball, $60,000,000;

b. MegaMillions, $39,000,000;

c. Megabucks Doubler, $1,100,000

10. a. $0.1 billion because 0.1 billion is 100,000,000, which is larger than 25,000,000 (25 million);

b. $1.5 billion because it is 1,500,000,000, while 150 million is 150,000,000.

c.

					whole					fr	
billions			1,000,000's millions	100,000's hundred thousands	10,000's ten thousands	1,000's one thousands	100's hundreds	10's tens	1's ones	$\frac{1}{10}$'s tenths	$\frac{1}{100}$'s hundredths
		2	5	0	0	0	0	0	0		
	1	0	0	0	0	0	0	0	0		
1	5	0	0	0	0	0	0	0	0		
	1	5	0	0	0	0	0	0	0		

Students may notice that each place has a value 10 times the place to its right, and a tenth the place to its left.

Practice: Grid Visions

Part 1

1. 0.1 of 10 = 1

2. 0.1 of 20 = 2

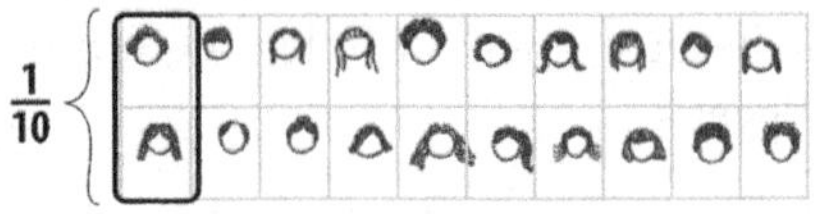

3. 0.1 of 100 = 10

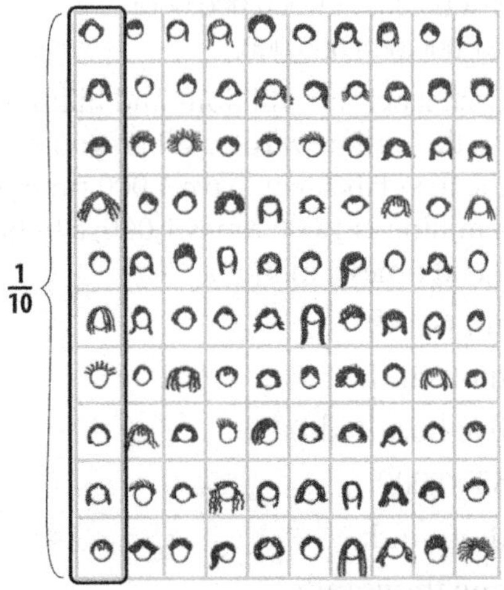

Reprinted with permission of *World Education*

Part 2

1. 0.1 of 30 = 3
2. 0.1 of 50 = 5
3. 0.1 of 120 = 12
4. Answers will vary. Students might notice they are using the same digits, but in a different place value.
5. Answers will vary. Possible answers: I adjust the place value of the digits; 50 becomes 5.
6. Answers will vary. Possible answer: I multiply the quantity by .1. 50 × .1 is 5.; I divide the quantity by 10.

Test Practice

1. (b)
2. (a)
3. (d)
4. (c)
5. (b)
6. 26 men

Lesson 3: What Is Your Plan?

Activity 1: 20% Dilemma

1. Answers will vary. Sample grid below:

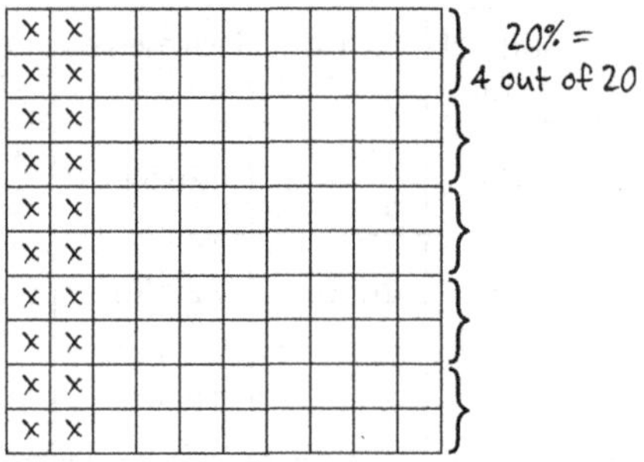

2. Answers will vary. Sample answers: 1/20; 5/100; 10/200

Activity 2: Display Plans

1. a. Answers will vary. Sample answer:

1 ft. (width and height of each square)

A	A	A	A	P	P	P	P	P	J
A	A	A	A	P	P	P	P	P	J
A	A	A	A	P	P	P	P	P	J
A	A	A	A	G	G	G	G	G	J
A	A	A	A	G	G	G	G	G	J

A = apples P = pears
G = grapes J = jams

2. 10 sq. ft. Explanations will vary. Possible explanations: Because 20% of 50 sq. ft. is 10 sq. ft.; the grapes are arranged in a 2 foot by five foot rectangle. 2' × 5' is 10 square feet; or if 10% of 50 is 5, then 20% is 10; or since 20% is 1/5 I knew to color in 1 out of every 5 squares, which would be 10 squares.
3. 15 sq. ft. Answers will vary. Possible explanations: Because 30% of 50 sq. ft. is 15 sq. ft. since 10% of 50 is 5, and 30% of 50 is 15; or the pears are arranged in a 3 foot by 5 foot rectangle. 3' × 5' is 15 square feet.

Activity 3: What Is *Your* Plan?

Answers will vary. Sample set of answers based on graphic below:

← 1 ft. → (1 ft. ↕)

T	T	T	T	T	T	T	T	T	T
T	T	T	T	T	T	T	T	T	T
T	T	T	T	T	C	C	C	C	C
K	K	M	M	M	M	M	C	C	C
K	K	K	M	M	M	M	M	C	C

K = key chains M = magnets
C = caps T = T-shirts

1. Our plan used 50 squares.
2. Item A: K = key chains; Item B: C = caps; Item C: M = magnets; Item D: T = T-shirts
3. 100%; 25 sq. ft + 10 sq. ft + 10 sq. ft + 5 sq. ft = 50 sq. ft.
4. a. Item A (key chains): 10%
   5/50 (1/10) of squares
   b. Item B (caps): 20%
   10/50 (1/5) of squares
   c. Item C (magnets): 20%
   10/50 (1/5) of squares
   d. Item D (T-shirts): 50%
   25/50 (1/2) of squares

Activity 4: Here Is 10%—What Is the Whole?

1. 10% Now
   a. Answers will vary. Possible answer:
   8 = 10%
   b. Answers will vary. Possible answer:
   80 = 100%
   c. Answers will vary. Possible answer:

8	8	8	8	8
8	8	8	8	8

2. Answers will vary.
3. Answers will vary. The last column values are ten times greater than the first column values.
4. Answers will vary. Possible answer: To find the whole when you know the part that equals 10%, multiply the part by 10.
5. Answers will vary. Possible answers: Not much, very little, closer to 0 than 1/2.

Practice: Controlling Costs for Seniors

1. a. 2
   b. 2
   c. 3
   d. 1
2. a. 2
   b. 2

Practice: Money Down

1. a. $50,000
   b. $100,000
   c. $120,000
2.

Person	Total Cost	Down Payment of 10%	Balance
a. Ella	$650	**$65**	**$585**
b. Irv	$3,300	**$330**	**$2,970**
c. Bella	**$1,900**	$190	**$1,710**
d. Burt	**$750**	$75	**$675**
e. Lou	**$500**	**$50**	$450

Practice: Drugstore Markups and Markdowns

1. a. $1.10
   b. $4.50
   c. $0.45
   d. $2.30
   e. $3.29
   f. $3.41
   g. $3.59
   h. $4.35
2. The cashier is right. 10% of the total of $15.74 = $1.57 for a total cost of $14.17; Or 10% of $3.50 = $0.35 for a cost of $3.15; 10% of $4.99 = $0.50 for a cost of $4.49; 10% of $1.25 = $0.13 (or $0.12) for a cost of $1.12 (or $1.13); 10% of $6.00 = $0.60 for a cost of $5.40. $3.15 + $4.49 + $1.12 (or $1.13) + $5.40 = $14.16 (or $14.17).

Practice: More Plans

Answers will vary. Sample answers based on a floor plan for an office.

Floor Plan

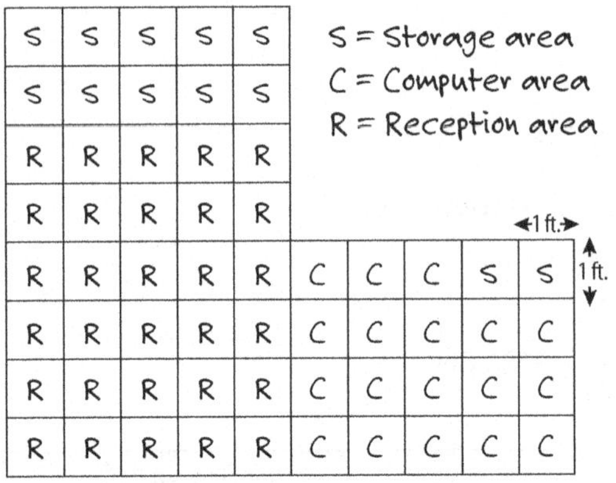

1. 20% is storage area; 30% is computer area; 50% is reception area
2. S = Storage; C = Computer; R = Reception
3. 10% of 60 is 6, so 20% is twice that amount: 12. 30% is 3 x 6, or 18. Half of 60 is 30, which is the same as 50%.
4. 21. If 10% of 60 is 6, then 30% is 6 x 3 = 18. 5% of 60 is half of 10%, or 3. 18 + 3 = 21.

Practice: Visual Percents

1. a. 75%
   b. 10%
   c. 60%
   d. 20%
   e. 90%
2. 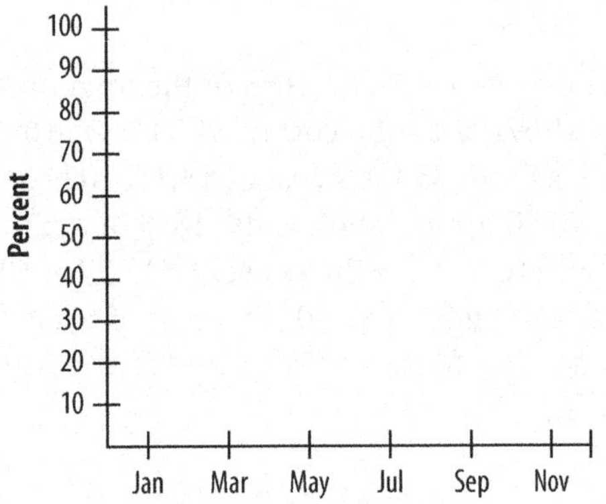

3. a. A drawing with 18 more stick figures, 20 in all.
   b. A drawing with 20 more stamps, 25 in all. Explanations will vary.
   c. A drawing with 7 more $5 bills, 10 in all. Explanations will vary.

d. 10

4.

a. 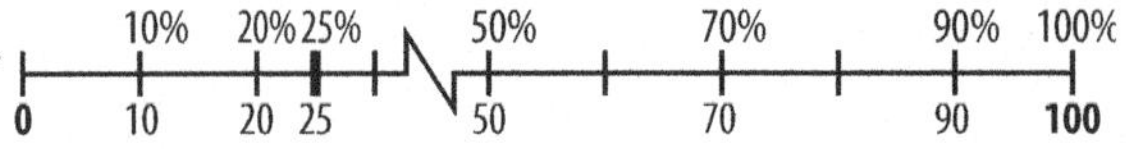

b. 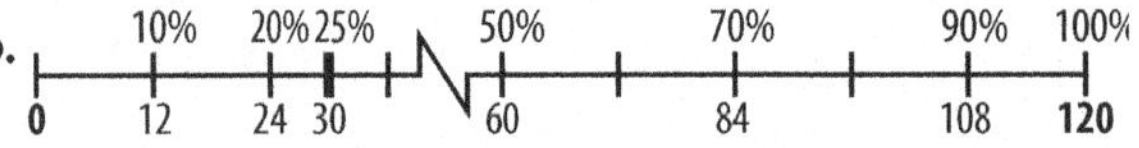

c. 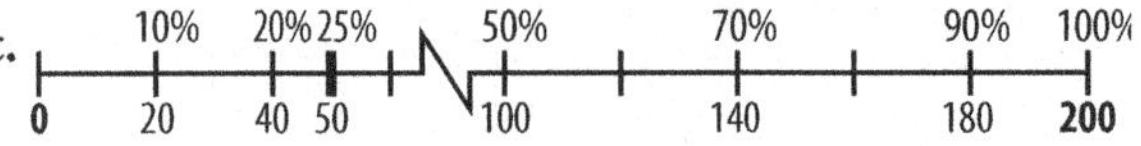

d. 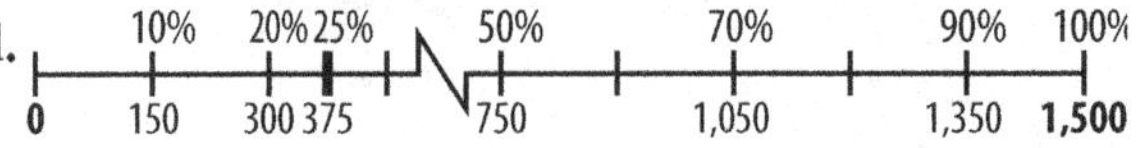

5. a. Percent shaded: 100%
   Fraction shaded: 100/100
   b. Percent shaded: 25%
   Fraction shaded: 25/100, or 1/4
   c. Percent shaded: 50%
   Fraction shaded: 50/100, or 1/2
   d. Percent shaded: 10%
   Fraction shaded: 1/10
   e. Percent shaded: 10%
   Fraction shaded: 5/50, or 1/10

Practice: Comparing Percents

1. 14 miles < 80%
2. 20 questions< 70%
3. 25 pennies> 40%
4. =
5. >
6. <
7. >
8. a. Answers will vary. Possible answer: 15 out of 150 is 10%, so 30 out of 150 is 20%.
   b. Possible answer for Problem 6: 120 out of 1,200 is 10%, 600 is 50% of 1,200, so 600 (50%) + 120 (10%) + 120 (10%) = 840 (70%). 800 is less than 840, therefore less than 70%.

Extension: Rami's Interest

1. (d) around 80 months. As the amount owed gets smaller, so does the amount for 10%. A spreadsheet may be used to see how this works out, but students may be able to reason this out by trying some of it with a calculator and noticing the gaps between the values for the 10% narrowing.

2. a.

	Total owed	Rami pays 10%	Amt Rami owes	Amt new interest charged
Jan	$100	– $10 =	$90	+ $9
Feb	$ 99	– $9.90 =	$89.90	+ $8.91
Mar	$98.01	– $9.80	$88.21	+ $8.82
Apr	$97.03	– $9.70	$87.33	+ $8.73
May	$96.06	– $9.61	$86.45	+ $8.65
June	$95.10	– $9.51	$85.59	+ $8.56
July	$94.15	– $9.42	$84.73	+ $8.47
Aug	$93.20	– $9.32	$83.88	+ $8.39
Sept	$92.27	– $9.23	$83.04	+ $8.30
Oct	$91.34	– $9.13	$82.21	+ $8.22
Nov	$90.43	– $9.04	$81.39	+ $8.14
Dec	$89.53	– $8.95	$80.58	+ $8.06

b. When Rami pays 10%, the amount he owes decreases. Then 10% of the decreased amount is charged, bringing his total back up nearly to what it was. This method of payment is ineffective.

c.

	Total owed	Rami pays 20%	Amt Rami owes	Amt new interest charged
Jan	$100	$20	$80	$8
Feb	$ 88	$17.60	$70.40	$7.04
Mar	$77.44	$15.49	$61.95	$6.20
Apr	$68.15	$13.63	$54.52	$5.45
May	$59.97	$11.99	$47.98	$4.80
June	$52.78	$10.56	$42.22	$4.22
July	$46.44	$9.29	$37.15	$3.72
Aug	$40.87	$8.17	$32.70	$3.27
Sept	$35.97	$7.19	$28.78	$2.88
Oct	$31.66	$6.33	$25.33	$2.53
Nov	$27.86	$5.57	$22.29	$2.23
Dec	$24.52	$4.90	$19.62	$1.96

d. This method (paying off at a higher percentage rate) is more effective and compared to 2a the total drops more quickly. However, as the amount owed drops, so does the amount of 20% he's paying. The gaps between values narrow, slowing the process as the months pass.

Test Practice

1. (d)
2. (d)
3. (a)
4. (d)
5. (b)
6. $150

Lesson 4: One Percent of What?

Activity 1: Go Figure!

Juan is right because it depends on what the whole is. In both cases the discount is $10. Ten percent of $100 is $10. Ten percent of $1,000 is $100, so 1% of $1,000 is $10.

Activity 2: Patterns with 10% and 1%

Answers will vary.

1. The decimal moved one place to the left (or in a whole number, a 0 was dropped each time) or the place value of the digits shifted one place of less value each time.
2. Answers will vary but should reflect an understanding that 10% is arrived at by dividing the amount by 10; 1% is arrived at by dividing the amount by 100.

Practice: Which Is Greater

1.

Left Column Amounts	Right Column Amounts
a. 12 steps ✓	10 steps
b. 30 voters	40 voters ✓
c. 27 million workers ✓	25 million workers
d. 11.5 pages	75 pages ✓
e. 1.5 games ✓	1 games
f. 4.5 TV shows ✓	3.5 shows
g. 100 viewers	150 viewers ✓
h. 38.5 strikers	39 strikers ✓

2. Possible answers: I thought of the numbers as shifting over one place value, making them less than they are in the original number; I imagined the number without the 0 at the end; I pictured the decimal point as being in its place at the end of the original number, then moving one place to the left to make the number 10 times smaller.
3. Possible answers: I thought of the numbers as shifting over two places less than they are in the original number; I pictured the decimal point as being in its place at the end of the original number, then moving two places to the left to make the number 100 times smaller.
4. Answers will vary. Students may describe moving the decimals to compare, or comparing the original value with the 1% with 10 times the value with the 10% which can be done by simply adding a zero to the end of the original amount for 10%.

Practice: Fundraisers

1.

Bake Sale	$ 800	$ 8	$ 80	$ 712
Car Wash	$ 600	$ 6	$ 60	$ 534
Raffle	$3,150	$31.50	$315	$2,803.50
Bumper Sticker Sales	$ 250	$ 2.50	$ 25	$222.50
Craft Fair	$1,200	$12	$120	$1,068
Walk-a-thon	$1,800	$18	$180	$1,602
Total	$7,800	$78	$780	$6,942

2. $78
3. 89%
4. By checking to see whether the totals of the columns are 1%, 10%, and 89% of $7,800, respectively. Also, by checking that the total of the gross income less the total of emergency funds and fundraiser costs equals the total remaining receipts.

Practice: Growing Cities and Towns

1.

Amador	40,000	400	40,400
Del Norte	28,000	280	28,280
Glenn	29,000	290	29,290
Mono	15,000	150	15,150
San Diego	3,272,000	32,720	3,304,720
San Fran	3,313,000	33,130	3,346,130
Sierra	3,000	30	3,030

2. Answers will vary. Possible answers refer to population in schools, availability of housing and other resources, and tax revenue for the town.

Practice: Rounding to the Nearest Whole Number

Part 1

1. b. or c. If b, there would be a little less meat and if c, there would be a little more meat. Each choice would be close to 2 pounds. "d" could be chosen if the explanation included something like freezing (2)6 1/2-ounce patties for another purpose.
2. **c.** Because Chris needs at least 1 1/2 (or 1.5) yards of cloth, 1.8 is the closest amount that isn't less than what she needs.
3. **c.** Since 5.3 million is the same as 5,300,000, it is larger than 5 million.
4. **b.** Since 7.6 is greater than 7.5 (or 7 1/2), it is closer to 8 pounds than to 7 pounds.

Part 2

1.

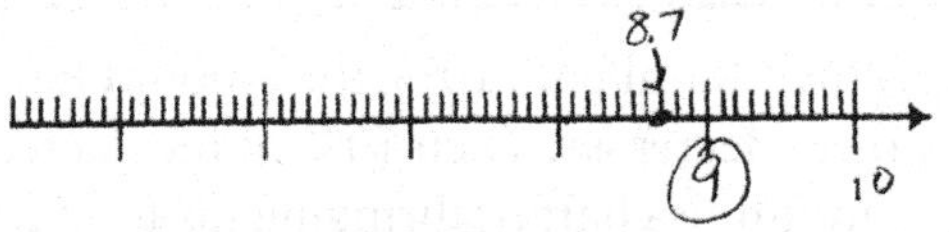

2.

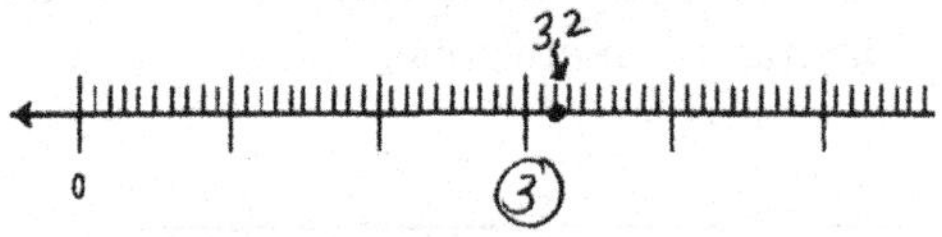

3.

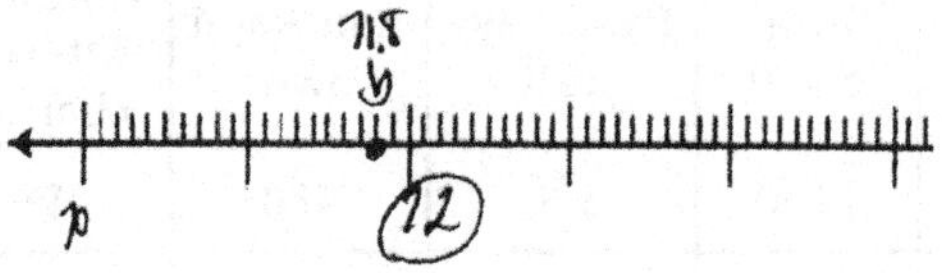

4.

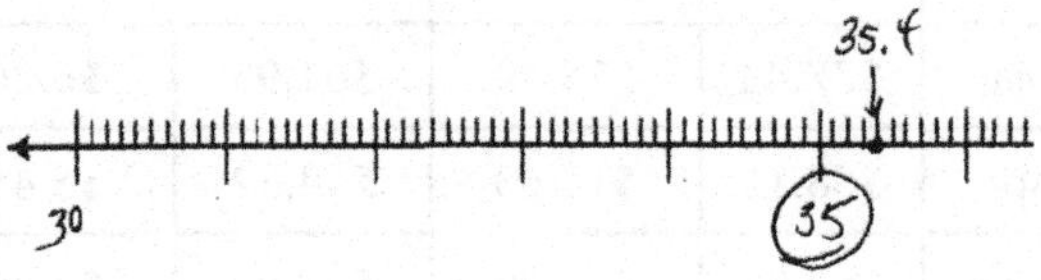

5.

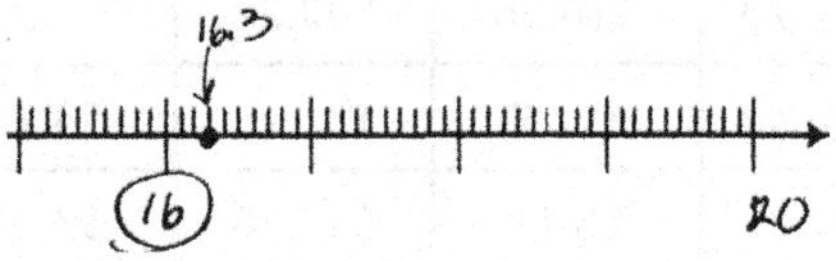

6.

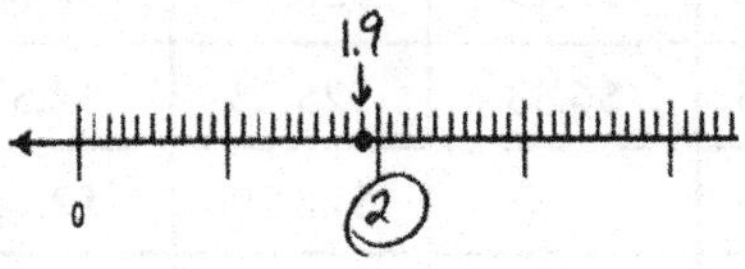

7.

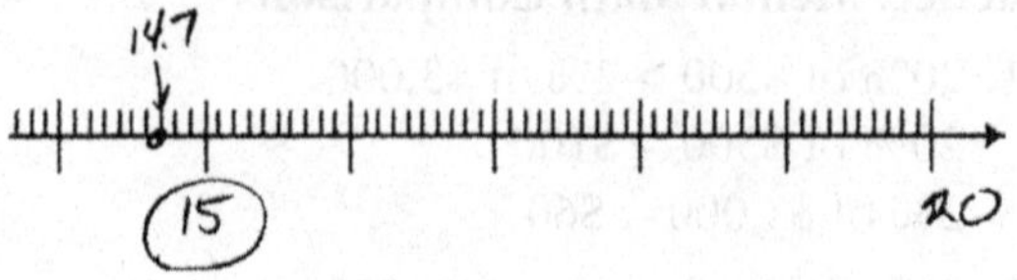

8.

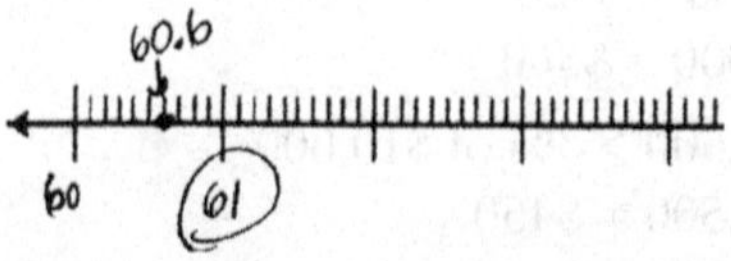

Extension: Working Backward

	1%	10%	100%
1.	2	200	2,000
2.	57	570	5,700
3.	18	180	1,800
4.	1.5	15	150

Test Practice

1. (b)
2. (e)
3. (c)
4. (b)
5. (d)
6. 333,700,000

Lesson 5: Taxes, Taxes, Taxes

Activity 1: Different States, Different Charges

Sales Tax Rate	Sound System $140	Sneakers $75	Washing machine $568
4%	$5.60	$3.00	$22.72
5%	$7	$3.75	$28.40
6%	$8.40	$4.50	$34.08
7%	$9.80	$5.25	$39.76

1. Possible answers: Find 10% and then take half of that amount; Find 1% and multiply that amount by 5 to find the value of 5%.
2. $148.40. The price is $139.99. The sales tax of 6% is $8.40. The total of those two costs is $148.40.
3. $2.80. The cost for the sound system is the same in both states, but in AL, I'll pay $5.60 in tax and in MI, I'll pay $8.40, which is $2.80 more.
4. The difference will be the difference in the amount of tax paid. $39.76 – $22.77 is $17.04.

Activity 2: Take-Home Pay

1. No. Mara will take home 54% of $1,600, or $864. Her brother will take home 72% of $400, or $288. Mara's take-home pay is three times her brother's.
2. $864
3. $288
4. Answers will vary but should include the idea that Mara pays $736 in taxes per week while her brother pays $112. Thus, Mara's take-home pay of $864 is three times as much as her brother's ($288).

Practice: Personal Payroll Deductions

Answers will vary. Sample answer based on $480 weekly gross for a single person in a state with an 8% income tax.

1. 40 hours x $12 = $480
2. 15%
3. $72.00
4. 8%
5. $38.40
6. 7%
7. $33.60
8. $336.00

Practice: Which Is a Better Deal?

1. Barkers: $50
   Elsa's: $45—better deal
2. Barkers: $34 + $2.04 tax = $36.04—better deal
   Elsa's: $40.28 (20% of $50 = $10, so sale price is $40; 5% of $40 is $2, so new sale price is $38 + $2.28 tax = $40.28). Some students may not calculate tax as another valid answer is that if it's a better deal before tax and tax is the same at each store, it would be a better deal after.
3. Barkers: $53.64
   Elsa's: $39.95—better deal
4. Barkers: $88—better deal
   Elsa's: $90
5. Barkers: $24.50 + $1.47 tax = $25.97—better deal
   Elsa's: $27.56 (60% of $65 = $39. $65 – $39 = $26.) 6% of $26 is $1.56, so new sale price is $26 + $1.56 tax = $27.56 Some students may not calculate tax as another valid answer is that if it's a better deal before tax and tax is the same at each store, it would be a better deal after.

6. Answers will vary.
Sample answer based on 1/5 off at Barkers and 25% off at Elsa's. Barkers: $3.92
Elsa's: $3.38—better deal

Practice: Increase, Decrease

1. 4,050
2. 1,104
3. 9,520
4. 1,160
5. 8,610
6. 731.5
7. a. $1,050 is 10% of $10,500.
$105 is 1% of $10,500.
   b. She should have subtracted $210 ($105 and $105) from the total of $2,100 ($1,050 + $1,050), not added them together. 18% = 20% – 2%.

Practice: Visuals of Percents

1. a. Entire circle is shaded.
   b. All of circle is shaded.
   c. 73%
2. 20. Sample answer: Each dot represents 5%, 20 x 5% = 100%
3. a. 150
   b. 180
   c. 150,000
   d. Answers will vary. Possible answer: if you figure out 10%, double it for 20%, and then find the middle of the two, you get 15%.
4. a. 65%
   b.

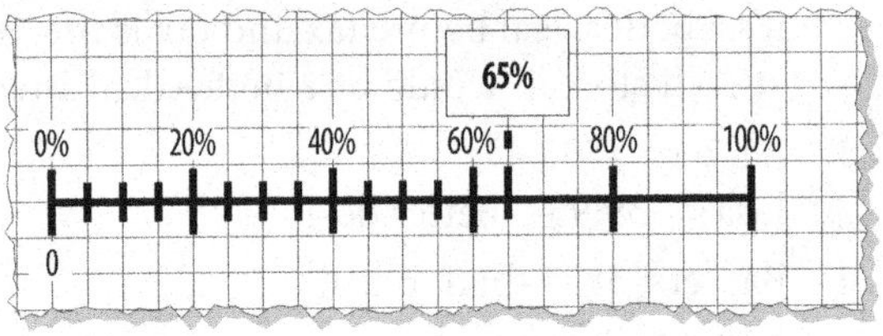

   c. 35%
   d. Answers will vary. Sample answer: 65% + 35% = 100%
5. a. A drawing with 140 blocks
   b. Answers will vary but should refer to the fact that if 35 blocks equal 25% of the whole, then 4(35) = 100%.

Practice: Mental Math Comparisons

1. 20% of $500 > 2% of $3,000
20% of $500 = $100
2% of $3,000 = $60
2. 40% of $800 < 4% of $9,000
40% of $800 = $320
4% of $9,000 = $360
3. 30% of $1,500 > 3% of $10,000
30% of $1,500 = $450
3% of $10,000 = $300
4. 60% of $200 = 6% of $2,000
60% of $200 = $120
6% of $2,000 = $120
5. 50% of $900 < 5% of $10,000
50% of $900 = $450
5% of $10,000 = $500
6. 70% of $1,200 = 7% of $12,000
70% of $1,200 = $840
7% of $12,000 = $840

Extension: Markdowns

1. Jacket, $125.00
$5
$120.00
$42.00
$78.00
$9.36
$68.64

   Lamp, $44.00
$1.76
$42.24
$14.78
$27.46
$3.30
$24.16
2. The 51% markdown off the original price is a better deal. Fifty-one percent of $125 is $63.75, resulting in a final sale price of $61.25. Adding up the percentages results in taking them all off the original price (a larger whole), giving a larger resulting discount.

Test Practice

1. (b)
2. (d)
3. (c)
4. (e)
5. (b)
6. 4,656

Lesson 6: Decimal Hundredths

Activity 1: Building on Fraction Strips

Part 1

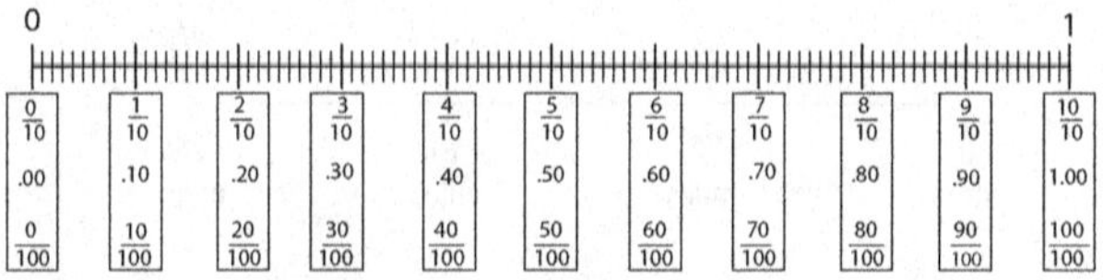

Part 2

1. Labels are (from the bottom):
   1/2= 0.50; 1/3 = 0.33, 2/3 = 0.67; 1/4 = 0.25, 2/4 = 0.50, 3/4 = 0.75;
   1/5 = 0.20, 2/5 = 0.40, 3/5 = 0.60, 4/5 = 0.80;
   1/6 = 0.17, 2/6 = 0.33, 3/6 = 0.50, 4/6 = 0.67, 5/6 = 0.83;
   1/8 = 0.125, 2/8 = 0.25, 3/8 = 0.375, 4/8 = 0.50, 5/8 = 0.625, 6/8 = 0.75, 7/8 = 0.875;
   1/12 = 0.08, 2/12 = 0.17, 3/12 = 0.25, 4/12 = 0.33, 5/12 = 0.42, 6/12 = 0.50, 7/12 = 0.58, 8/12=0.67, 9/12 = 0.75, 10/12 = 0.83, 11/12 = 0.92;
   1/10 = 0.10, 2/10 = 0.20, 3/10 = 0.30, 4/10 = 0.40, 5/10 = 0.50, 6/10 = 0.60, 7/10 = 0.70, 8/10 = 0.80, 9/10 = 0.90.
2. **a.** 1/8 = 0.125 or 1/12 = 0.08 are both close to 0.10; 1/8 is more than 0.10 while 1/12 is less than 0.10
   b. 1/3 or 0.33 is close to but more than 0.30; 1/4 or 0.25 is close to but less than 0.30.
   c. 11/12 or 0.92 is close to but more than 0.90; 7/8 or 0.88 is close to but less than 0.90.
3. **a.** True; 0.10 is 1/10 which is less than 1.
   b. True; 0.25 and 1/4 are two names for the same value.
   c. True; 1/3 is 0.33 which is larger than 0.25, or 1/3 is more than 1/4.
   d. False; 1/8 is 0.125 which is larger than 0.10, or 1/8 is more than 1/10.
   e. False; 1/3 is 0.33 which is larger than 0.15.
   f. True; 1/6 is about 0.17 which is smaller than 0.20, or 1/6 is less than 1/5.

Part 3

You would use 0 in the hundredths place for money, and also to indicate that in measuring something to hundredths, you haven't rounded to the tenths place value.

Activity 2: Extending the Places to Hundredths

Part 1

1. They are all right. Explanations should include reference to the fact that 7 maintains a place value in the tenths place even when it is called 70 hundredths—they are all different names for the same value.
2. ten thousands, thousands, hundreds, tens, ones. tenths, hundredths

Part 2

0.09, .9, 1.09, 1.8, 1.90, 8.00, 9.01, 09.09, 18

23.43, 23.44, 23.45, 23.46

23.61, 23.71, 23.81, 23.91

Activity 3: Finding the Match

1. Answers will vary for both **a.** and **b.** Students could match all percents, or all decimals, or all fractions or they could match equivalencies such as 80%, 0.80, 80/100.
2. **a.-c.**

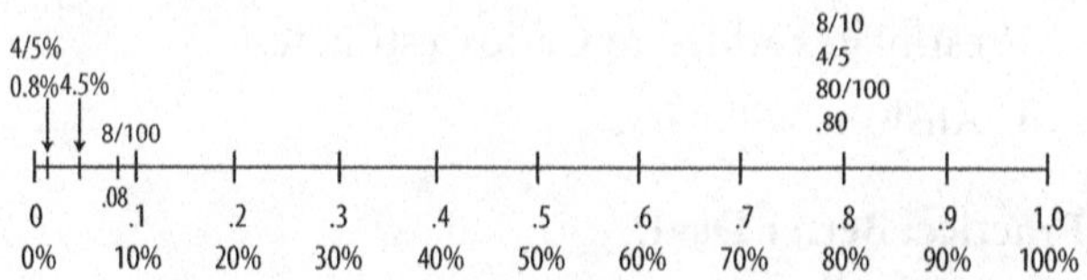

   d. There are many equivalences; 80% = 80/100 = .80 = 8/10 = 4/5, for example.
   e. While fractions, decimals, and percents may look different, they are equivalent when they occupy the same place on the number line.

Activity 4: Runners' Dilemma

1. Manny rounded each time to the tenths place; 9.75 = 9.8, 9.79 = 9.8 and 9.81 = 9.8.
2. Luis organized the data by looking at the values in the hundredths place.
3. Don looked at the tenths place without considering rounding; therefore to Don 9.75 and 9.79 are 9.7 while 9.81 and 9.88 are 9.8.
4. Of all the decisions, Luis is using good number sense.

Practice: Designer Accessories

1. 36 P's; 36 J's, 12 H's; 12 B's
2. P = Purses; J = Jewelry; H = Hats and Scarves; B = Belts
3. 12. 1/8 of 96 is 12.
4. 36. 3/8 of 96 is 36.

Practice: Round Up or Down?

1.

Month	Total Sales	Carlos' 12% Estimate for Retirement	Sonia's 13% Estimate for Retirement
1	$ 1,000	$ 120	$ 130
2	$ 800	$ 96	$ 104
3	$ 6,500	$ 780	$ 845
4	$ 4,000	$ 480	$ 520
5	$ 1,600	$ 192	$ 208
6	$10,000	$1,200	$1,300
	$23,900	$2,868	$3,107

2. a. Carlos = $2,868

   b. Sonia = $3,107

3. $2,987.50 (12.5% of $23,900 = $2,987.50) or half the difference between Carlos' and Sonia's estimates added to Carlos' estimate

4. Answers will vary.

Practice: Better Deal?

Part 1: The Facts

1. Sample answer: To find 1/3, divide the amount by 3.
2. Sample answer: To find 2/3, divide the amount by 3, then multiply by 2.
3. Sample answer: To find 30%, find 10%, then multiply by 3.
4. 1/3 = 33 1/3%
5. a.

Part 2: Which Deal Is Better?

Students should have circled the item they thought was the better deal before they calculated.

1. Barkers: $42
   Elsa's: $40—better deal
2. Barkers: $28—better deal
   Elsa's: $33.50
3. Barkers: $24
   Elsa's: $18.36—better deal
4. Barkers: $73.70—better deal
   Elsa's: $78
5. Barkers: $23.10—better deal
   Elsa's: $26

Practice: How Much, How Far?

1. a. No

   Possible answer: Val should walk 8 miles to walk 1/3 of the way.

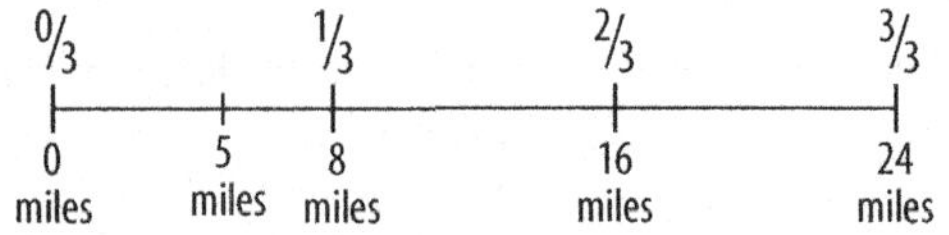

   b. 66.66%, or $2,000

   c. 100%, or $3,000

   d. Since I knew that Val's portion was 1/3 of the money, I doubled it to find Solé's portion, 2/3. I added Val's portion and Solé's portions to get the whole. I did the same thing with the percents.

2. a. 14th

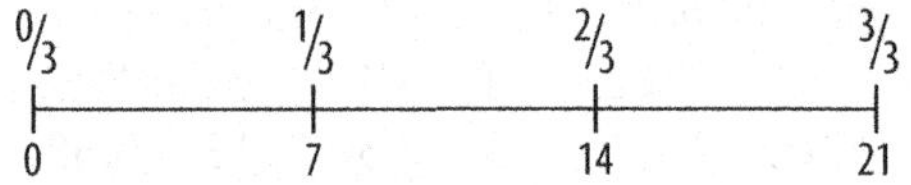

   b. 1/3

   c. More. Sample explanation: One-third of an amount is always more than one-fourth of the same amount.

Extension: Budgeting

1. Answers will vary, but total dollar amount should equal $2,100, and total fraction and percents should equal 1, or 100%.
2. Answers will vary, but total dollar amount should equal the amount listed under monthly income. Fraction and percents should equal 1, or 100%.

Test Practice

1. (d)
2. (e)
3. (d)
4. (e)
5. (b)
6. Answers will vary and may include 1.21, 1.22, . . . , 1.29.

Lesson 7—Smaller and Smaller

Activity 1: Target 1, 1/2, and 1/4

1. Answers will vary. The closest option could have a number in the ones place and other two digits as decimals, or a decimal followed by the highest number in the tenths place.
2. Answers will vary. Look for answers with the three digits to the right of the decimal point forming a decimal close to ½ or a one to the left of the decimal with a small digit in the tenths place.
3. Answers will vary. Again the decimal point should precede the three numbers and form a decimal close to 0.25.
4. Answers will be different depending on what the student chose.
5. Answers will vary. For the second and third questions, students might notice that starting with the tenths place and a number closest to the decimal equivalent for the fraction gets them in proximity to the target (e.g., of 5-2-7, starting with .5 for a target of one-half; .2 for a target of one-quarter, and .7, the largest digit, to get closest to the target of 1).

Activity 2: Smaller and Smaller

Part 1: Less Than a Penny

1. Card A: $0.113

   Card E: 1/2 cent

   Card K: 3.37 9/10, 3.58 9/10, 3.77 9/10

   Card N: All the prices are written to the thousandth of a dollar, or tenths of a cent.
2. Card A: In $0.113, the 3 is less than 1 cent. It is worth 3/10 of a cent, or 3/1000 of a dollar.

   Card E: The 1/2 cent is less than 1 cent. It is worth .5 of a cent or .005 of a dollar. With a dollar sign, it can be written as $0.005.

   Card K: In 3.37 9/10, 3.58 9/10, 3.77 9/10, the 9/10 is less than 1 cent. It is worth .9 of a cent or .009 of a dollar. With a dollar sign, it can be written as $0.009.

   Card N: All the prices are written to the thousandth of a dollar, or tenths of a cent. Thus the digit in the thousandths place represents that many thousandths of a dollar or tenths of a cent.

Part 2: Less than 1%

1. Card H 62.5%, Card L 0.08%, Card M 2.5%, Card N 3.5%. All the percent change amounts are written to the nearest tenth of a percent.
2. a. Card L 0.08% < 1%. Card N, -.215 and -.205< 1%

   b. For Card L: 0.08% = .0008 and .1% = .001. For Card N: Any percent that is negative will be less than 1%, although in this case, the negative does not represent less than zero, but rather a percent change.

   c. Card H: 0.625; Card L: 0.0008; Card M: 0.025; Card N: .197, .046, and .215 all the percents are greater than 1% (except for the negative percents on Card N).

Part 3: Picturing Thousandths

1. Cards B , D, and F
2. Card B: The larger intervals are worth .1; the smaller intervals are worth 1/10 of .1, or .01. The thousandths are not marked—they are so small.

   Card D: The large flat has a value of 1. The rod has a value of .1, and there are 10 rods (.1) to a flat (1). The square has a value of .01, and there are 10 squares (.01) to a rod (.1). The small piece has a value of .001, and there are 10 small pieces (.001)to a square (.01).

   Card F: The one flat represents 1, the 3 ten rods represent 3 tenths, the 7 squares represent 7 hundredths, and the small piece represents 5 thousandths, or a half of a hundredth.
3. a. 1 inch

   b. 0.1 inch

   c. 0.01 inch

   d. 0.001 inch

Activity 3: Expanded Notation

Answers will vary here depending on what values the student chooses. For example:

0.113 1(1/10) + 1(1/100) + 3(1/1,000)

3.521 3(1) + 5(1/10)+ 2(1/100) + 1(1/1,000)

3.14159 3(1) + 1(1/10) + 4(1/100) + 1(1/1,000) + 5(1/10,000) + 9(1/100,000)

Activity 4: Eight Questions

1. 3.14; Depending on which website they choose, pi's decimal places are infinite because pi is an irrational number. Irrational numbers can be represented as decimals that never repeat and never end. However a computer computed pi to 5 trillion places on October 18, 2011.
2. Using the width of a piece of hair ranging from 0.003 to 0.007 of an inch thick, the range of answers is approximately 143 hairs to 333 hairs. If students have reasonable totals based on different hair widths, from their own internet research, accept them as well.
3. .125, .250, .375, .500, .625, .750, .875, 1.000, 1.125
4. 12.5%, 25%,37.5%, 50%, 62.5%, 75%, 87.5%, 100%, 112.5%
5. At the end of the 2012 season, Alex Rodriquez's lifetime average was .300. As a percent this would be 30% and as a fraction 3/10.
6. Batting 1000 means having a hit at every bat or 1000/1000 = 1.000 = 1. Batting 500 means .500 which is 1 hit for 2 at bats.
7. Given that 2/3 is a repeating decimal the calculator read out will be 0.66666667 (the calculator rounds the last digit).
8. All states have set the legal BAC limit at 0.08%, which is less than 1/10 of 1%.

Practice: Three Decimal Places

1. Hundred thousands, ten thousands, thousands, hundreds, tens, ones, decimal point, tenths, hundredths, thousandths
2. .009, 0.09, 1.090, 1.90, 8.00, 9.015, 9.09, 18
3. They are all correct, but should explain that the place value is the same, but these are different ways of describing it.
4. 9.5, 9.6, 9.7, 9.8, 9.9
5. 9.5, 9.51, 9.52, 9.53, 9.54
6. 9.5, 9.501, 9.502, 9.503, 9.504

Calculator Practice: Fraction-Decimal-Percent Conversion

Fraction	Decimal	Percent
1. 105/100	1.05	105%
2. 35/100 = 7/20	0.35	35%
3. 15/10 = 3/2	1.5	150%
4. 19/12	1.58333...	158.3%
5. 108/100 = 27/25	1.08	108%
6. 59/100	0.59	59%
7. 16/64	0.25	25%
8. 5/52	0.09615385	9.6%
9. 2/100 = 1/50	0.02	2%

10. Answers will vary.

Practice: Rounding to the Nearest Tenths and Hundredths

1.

	Rounded to nearest tenth	Rounded to nearest hundredth
a. $0.113	$0.1	$0.11
b. 0.452	0.5	0.45
c. 3.14159	3.1	3.14
d. 0.001	0.0	0.00
e. $0.005	$0.0	$0.01
f. 1.375	1.4	1.38
g. 0.333....	0.3	0.33
h. 0.625	0.6	0.63
i. .300	.3	.30
j. .007	.0	.01
k. 3.589	3.6	3.59

2. a. True
   b. False
   c. False
   d. True
   e. True
   f. False

3.

Year	Rounded price	Highest to lowest
2002	$2.81	$2.69
2003	$2.69	$2.81
2004	$2.88	$2.88
2005	$3.30	$3.07
2006	$3.20	$3.20
2007	$3.07	$3.24
2008	$3.87	$3.30
2009	$3.58	$3.30
2010	$3.24	$3.46
2011	$3.30	$3.58
2012	$3.58	$3.58
2013	$3.46	$3.69
2014	$3.69	$3.87

4. a. The median price is $3.30.
   b. No. The prices changed but did not necessarily increase every year.

Practice: Splitting Tips

1. a. 1% is 20¢ × 27 = .20 × 27 = $5.40
   b. 1/2 of $20 is $10 and half again is $5. Moe makes $5.
   c. 10% is $2.00; $2.00 × 2 is $4.00 and add half of 10% is $1, so $4 + $1 = $5
   d. $20 ÷ 8 = 2.50; $2.50 × 2 = $5.00
   e. 40¢
2. a. $1.50
   b 7.50
   c. $20.10
   d. Accept answers from $37.50-$40.50
   e. Accept answers from $37.50-$40.50
   f. Accept answers from $37.50-$40.50
   g. Answers will vary.
   h. 0.269 × 150 = $40.35
   i. Answers will vary.

Extension: Rate Hikes—Ouch!

1. $0.0025
2. a. $2.395 or $2.40
   b. $28.74
3. Answers will vary.
4. Answers will vary.

5. a. .00125 • 958 = $1.1975 or $1.20 per month and $14.37 per year
   b. .0001 • 958 =$0.0958 or $0.10, or 1.1496 or $1.15 per year

Test Practice

1. (c)
2. (e)
3. (d)
4. (d)
5. (b)
6. Answers will vary and may include 8.111, 8.112, … , 8.119

Lesson 8: Adding and Subtracting Decimals

Activity 1: Watch Out

Scenario 1

a. No. Anita's calculation for Lila is incorrect. Disregarding the decimal point and adding the time as if the amounts were all whole hours (2 + 1 + 4) gives Lila 7 hours. For Aman, Lila's math is correct.

b.

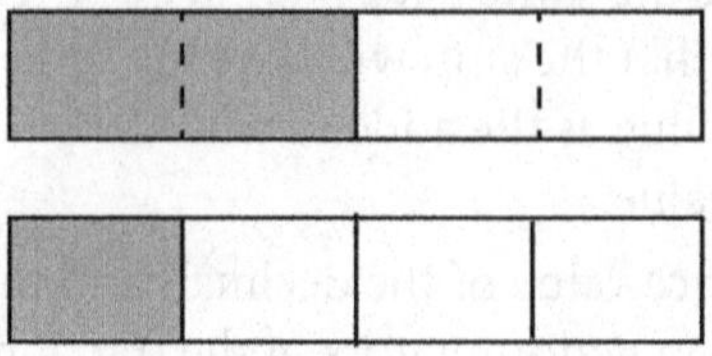

This shows that 1/2 + 1/4 = 3/4 for Lila, so 7 3/4 hours total. Then she needs to include the 6 1/4 hours for Aman, which gives a total of 14 hours.

c. .5 + .25 = .75 Lila's total is 7 3/4 hours

Anita added the hours without considering the decimal points (451), then considered the first number the number of hours and the second two numbers the number of minutes.

d. 14 hours total for the two volunteers

Scenario 2

No, she added the values without considering the decimal point. To estimate 0.8 would be close to 1 pound, 1.2 is close to one pound, and 0.4 pound is close to 0.5 pound, weighing around 2.5 pounds.

Scenario 3

No, Anne beat Harvey by 1.32 minutes (41.32 – 40.0 = 1.32). Mia beat Sam by 0.02 minutes (35.02 – 35.0 = 0.02). Lucy beat Mark by 0.03 minutes (46.03 – 46.0 = 0.03).

Activity 2: Methods for Adding and Subtracting Decimals

Set 1: Look for Patterns

1. Students may notice that "likes" are combined with "likes." For example in **a** and **b** the tenths are all combined, and the hundredths are also combined. Problem **c** shows that when adding fractions you don't add the denominator.
2. Answers will vary. Students may mention that to add or subtract you line up the decimal point. Also, that you combine or subtract likes from likes, and this is true for both fractions and decimals.

Set 2: Keep Looking for Patterns—Develop a Method

1. Students may notice that with decimals, if you add zeroes at the end of the number until all the numbers are to the same decimal place you can use the same pattern as in Set 1. They may notice that the sum will have the same smallest place value as the addend with the smallest place value.
2. The place value of the decimals and the common denominators of the fractions work the same way for addition and subtraction.

Set 3: Correct or Not?

1. & 2. Equations **a** and **b** are correct; **c** is not correct because the denominators were added rather than finding and keeping a common denominator; the answer is 1 1/6 or 1.6666; **d** is not correct because the decimals weren't lined up according to place value before adding; the answer is 3.554.

Activity 3: Tolerances

1. 32.40 cm to 33.10 cm (zeroes are necessary in this answer)
2. 0.095 mm to 0.105 mm
3. 56 5/8 inches to 57 1/8 inches
4. 39.6 psi to 40.4 psi
5. 4.70 mm

Practice: Where Is the Point?

Part 1

1. 4.226
2. 120.6
3. .302
4. 30.2
5. 3.02
6. Answers may vary but should include looking at place value and correctly adding likes to likes (tenths to tenths, hundredths to hundredths, etc.)
7. d

Part 2

1. 8.186
2. 126.6
3. .302
4. 30.2
5. 302.
6. b
7. c

Practice: Closest Estimate

1. 1/10
2. 1.25
3. 300/1,000
4. 1.5
5. 1
6. .5
7. .25
8. 9/10
9. 1.7
10. d

Calculator Practice: Fractions and Decimals

Problem	Your Estimate	Calculator Answer
1. 0.5 – 1/5	Possible answer: smaller than a half and close to a half of a half	1/5= 0.2 0.3
2. 1.75 + 1/8	Possible answer: a little less than 2	1.875
3. 0.08 + 2/3	Possible answer: a little more than 2/3	0.7467 rounded

4. 2/3 – 0.08	Possible answer: a little more than a half	2/3 = .67 0.59 rounded
5. 16.67 + 2 1/3	Possible answer: since 0.67 and 1/3 are 1 the answer should be close to 19	19.00333
6. 0.375 + 12/72	Possible answer: close to 0.5 since 12/72 = 1/6 and 0.375 is close to 1/3.	0.5417 rounded

Extension: Gone Fishing

1. 8.57 – 8.49 = 0.08 feet longer
2. **a.** 3 lbs. 11.787 oz.

   b. 38 lb. 12.64 oz

Test Practice

1. (d)
2. (b)
3. (c)
4. (b)
5. (d)
6. 128.081

Lesson 9: Multiplying Decimals

Activity 1: Show Me ...

1. a. Answers will vary. For example, a student could show 3 tenths strips, or show 1/10 of 3 on the number line, etc.

b. Answers will vary. For example, a student could say triple the measurement 0.1 cm.

2. a. Answers will vary. For example, a student could show 1.75 ten times.

b. Answers will vary. For example, a student could say they bought 10 candy bars that cost $1.75 each.

3. a. Answers will vary. For example, a student could show ½ of 12.5 on a number line.

b. Answers will vary. For example, someone bought 12.5 pounds of candy that sold for $0.50 a pound.

c. Sima's thinking is correct; 50%, 1/2, and 0.5 are all different ways to say half. Ron's 5% suggestion is equivalent to 5/100 or 1/20 and 0.05. Miguel's point about Ron's 5% is that it does not have the same value as half. When Miguel says, "They are decimals," he may be pointing out that the percent's place value must be considered, just as in the decimals they are replacing in the original problem.

Activity 2: Methods for Multiplying Decimals —They Have to Make Sense!

Part 1

1. (1) 3 × 1/10 = 3/10
(2) 3/10 = 0.3
(3) calculator 0.3
(4) Possible answer: 3/10 is the same as 0.3

2. (1) 1 3/4 × 10 = 17 1/2
(2) 17.5
(3) 17.5
(4) Possible answer: 17½ is the same as 17.5

3. (1) 12 1/2 × 1/2 = 6 1/4
(2) 6.25
(3) 6.25
(4) Possible answer: 6¼ is the same as 6.25

Part 2

Set 1: Find a Pattern

1. Some possible conversations may include: 3 sets of one-tenth; 1 3/4 of 10, so one 10 and 3/4 of 10; 4.8 groups of 906, so 4(906) + eight-tenths of a group of 906. These could be drawn.

2. Possible answer: When multiplying by a whole number, tenths give decimal tenths in answer, hundredths give decimal hundredths in answer, thousandths give decimal thousandths in answer.

3. 1) Multiply the digits. (2) When you multiply a whole number by a decimal with tenths, you will have tenths in your answer. When you mutiply a whole number by a decimal with hundredths, you will have hundredths in your answer. The pattern is true for thousandths as well. Place the decimal accordingly. (3) Check the reasonableness of your answer.

4. Students should find that their method works for all problems like those in Set A. Look and listen for connections between their visual solution and their steps.

Set 2. Keep Looking for Patterns

1. Some possible conversations may include: 12 1/2 pounds of bulk flour that costs $.50 per pound; 9 tenths of a walk that is 1.5 miles; 4.8 hours labor charged at $90.60 per hour.

2. They are all decimals to the tenths place multipled by other decimals to the tenths place. All the answers are to the hundredths place. The numbers being multiplied are a mix of numbers smaller than one and larger than one.
3. Possible answer: (1) Multiply the digits. (2) When you multiply tenths by tenths, your answer will be in hundredths; place the decimal there. (3) Check to see if your answer makes sense.
4. Students should find that their method works for all problems like those in Set B. Look and listen for connections between their visual solution and their steps.

Set 3: Develop a Method

1. They are all decimals multiplied by decimals. Some are tenths, some hundredths, and one thousandth.
2. (1) Multiply the digits. (2) The place value of the answer is the result of multiplying the place values in the numbers being multiplied. For example, tenths by tenths is hundredths, so the answer is hundredths. Tenths by hundredths is thousandths, so the answer is in the thousandths. (3) Check your answer for reasonableness. Students may also recall the commonly taught shortcut about the product having the same number of digits to the right of the decimal as the combined number of digits to the right of the decimals in the factors.
3. Students should find that their method works for all problems like those in Set 3. Look and listen for connections between their visual solution and their steps.

Activity 3: Calculators—What Happened?

1. 25/1,000 × 4/10 = 100/10,000 = 1/100 = 0.01 not 0.1. The decimal point was misplaced, perhaps by typing .25 × .4 or .025 × 4
2. 8/10 × 125/1,000 = 1,000/10,000 = 1/10 = 0.1 not 1. The decimal point was misplaced, perhaps by typing 8 × .125 or .8 × 1.25.
3. Just estimating you see that 2/1,000 × 7 would be 14/1,000 = .014 not 0.148. The decimal point was misplaced, perhaps by typing 0.02 × 7.4 or 0.002 × 74.
4. Just estimating you see that 35 × 1/100 = 35/100 = 0.35 not 3.56. The decimal point was misplaced, perhaps by typing 35.6 × 0.1 or 356 × 0.01.
5. Just estimating you see that 9 × 10,000 would be 90,000 not 95.9; perhaps the comma for 10,000 was mistaken for a decimal.

Activity 4: Shortcuts—Multiples of 10

1. Set 1
   a. 3.6
   b. 0.36
   c. 36
   d. 0.036
   e. The decimal point moves one place to the left from the original number being multiplied by 0.1, creating a smaller number.
2. Set 2
   a. 0.78
   b. 7.8
   c. 0.078
   d. 0.0078
   e. The decimal point moves two places to the left from the original number being multiplied by .01, creating a smaller number.
3. Set 3
   a. 9.4
   b. 0.94
   c. 94
   d. 940
   e. The decimal point moves one place to the right from the original number being multiplied by 10, creating a larger number.
4. Set 4
   a. 6700
   b. 670
   c. 6.7
   d. 67
   e. The decimal point moves two places to the right from the original number being multiplied by 100, creating a larger number.
5. Possible answer: When multiplying by 10, 100, or 1,000, the decimal point will move to the right according to the number of zeroes: one place for 10 (one zero in 10), two places for 100 (two zeroes in 100) and three places for 1,000 (three zeroes in 1,000).

Math Inspection: Exponents

Part 1: Rewriting using Exponents

Problem	Rewritten using exponents	Answer
3.5 x 3.5	3.5^2	12.25
7.6 x 7.6	7.6^2	57.76
.1 x .1	$.1^2$	.01
.01 x .01	$.01^2$	.0001
2.01 x 2.01	2.01^2	4.0401

Part 2: Patterns with Exponents

1. Set 1: Multiples of 10
   a. $10^4 = 10{,}000$
   b. $10^3 = 1{,}000$
   c. $10^2 = 100$
   d. $10^1 = 10$
   e. $10^0 = 1$
   f. $10^{-1} = 0.1$ or $1/10$
   g. $10^{-2} = 0.01$ or $1/100$
   h. $10^{-3} = 0.001$ or $1/1{,}000$
   i. $10^{-4} = 0.0001$ or $1/10{,}000$
2. a. It equals 1.
   b. It equals the number that was raised to the power of 1.
   c. The answer is less than 1.
   d. Answers may vary, but they might notice the number of zeroes in the whole numbers is the same as the power, or that the number of decimal places is the same as the negative power.
   e. Same: they are both powers of 10; different: their exponents are opposites of each other, the one with the positive exponent = 1,000 while $10^{-3} = 0.001$ or 1/1,000.)
   f. Same: they are both powers of 10; different: their exponents are opposites of each other, the one with the positive exponent = 10,000 while $10^{-4} = 0.0001$ or 1/10,000.
3. Set 2: Multiples of 2
   a. $2^4 = 16$
   b. $2^3 = 8$
   c. $2^2 = 4$
   d. $2^1 = 2$
   e. $2^0 = 1$
   f. $2^{-1} = 1/2$
   g. $2^{-2} = 1/4$
   h. $2^{-3} = 1/8$
   i. $2^{-4} = 1/16$
4. a. It equals 1.
   b. It equals the number that was raised to the power of 1.
   c. The answer is less than 1.
   d. Answers may vary, but they might notice that the negative powers are reciprocals of the positive powers.
   e. Same: they both are powers of 2; different: the negative power is the reciprocal of the positive power or the positive power is a whole number while the negative power is a fraction or decimal less than 1.
   f. Same: they both are powers of 2; different: the negative is the reciprocal of the positive power or the positive power is a whole number while the negative power is a fraction or decimal less than 1.
5. Set 3: Multiples of 1—You Try It!
   a. $1^4 = 1$
   b. $1^3 = 1$
   c. $1^2 = 1$
   d. $1^1 = 1$
   e. $1^0 = 1$
   f. $1^{-1} = 1$
   g. $1^{-2} = 1$
   h. $1^{-3} = 1$
6. All of the answers = 1 because 1 times itself = 1 and 1/1 = 1.
7. Set 4—Now You Try Another Number!

   Answers will vary depending on what number students pick.

   Same: they both had the same number raised to a power; different: the one with a positive exponent will be a whole number, while the one with a negative exponent is its reciprocal and is a fraction or decimal less than 1.
8. Same: the numbers are both powers of 3; different; the negative power is the reciprocal of the positive power or the positive power is a whole number while the negative power is a fraction or decimal less than 1.

Part 3: More Patterns

Problem	Expanded	Result	Another look
c. $10^2/10^3$	$\frac{10 \times 10}{10 \times 10 \times 10}$	1/10	2 – 3 = -1 and 10^{-1} = 1/10
d. $10^4/10^3$	$\frac{10 \times 10 \times 10 \times 10}{10 \times 10 \times 10}$	10	4 – 3 = 1 and 10^1 = 10
e. $10^3/10^1$	$\frac{10 \times 10 \times 10}{10}$	100	3 – 1 = 2 and 10^2 = 100
f. $10^6/10^6$	$\frac{10 \times 10 \times 10 \times 10 \times 10 \times 10}{10 \times 10 \times 10 \times 10 \times 10 \times 10}$	1	6 – 6 = 0 and 10^0 = 1
g. $10^4/10^2$	$\frac{10 \times 10 \times 10 \times 10}{10 \times 10}$	100	4 – 2 = 2 and 10^2 = 100

2. Possible patterns noticed: When multiplying with powers of 10, the answer is the same as adding the powers, and when dividing with powers of 10, the answer is the same as subtracting the denominator's exponent from the numerator's exponent.

Math Inspection: Break Apart the Numbers

1. a. 30

b. 18

c. 19

2. Dana's way
4.5 × 7 = 4(7) + .5(7) = 28 + 3.5 = 31.5

Lou's way
7 × .5 = 3.5

7 × 4 = 28

28 + 3.5 = 31.5

Dana's way
3.7 × 5 = 3(5) + .7(5) = 15 + 3.5 = 18.5

Lou's way
5 × .7 = 3.5

5 × 3 = 15

3.5 + 15 = 18.5

Dana's way
2.1 × 9 = 2(9) + .1(9) = 18 + 0.9 = 18.9

Lou's way
9 × .1 =0.9

9 × 2 = 18

0.9 + 18 = 18.9

3. Answers will vary. For example, they are the same; they keep one number whole and break the other apart by place. They both keep the group as the whole, and distribute the parts (as in the seven in 4.5 groups of 7). The method that is easier or faster will vary.

Math Inspection: Shortcuts with 0.5 and 0.25

Part 1: Which Method Is Correct?

All methods are correct. Possible picture could be a drawing of 36 split into 4 groups of 9.

Part 2: Describe the Rule

1. Set 1

a. 163

b. 163

c. 163

d. 39.3

e. 39.3

f. 39.3

2. Possible answers: Dividing by 2 is the same as multiplying by 1/2 or 0.5.

3. The answers would be the same, but they might be written differently. For example: 50% of 326.

4. Set 2

a. 1

b. 42

c. 40

d. 229

e. 66

f. 393

5. Possible answer: Students may see a relationship between 0.125, 0.25 and 0.5 and their fraction equivalents 1/8, 1/4, and 1/2. If they recognize that equivalence, then it would follow that the numbers multiplied are like dividing by 8 (0.125) or 4 (0.25) or 2 (0.5).

Part 3: Fill in the Blanks

a. 2

b. 4

c. 2

d. 1/4

e. 0.5

Practice: Draw It!

1. a. Yes; possible sketch:

1. b. Yes; possible sketch:

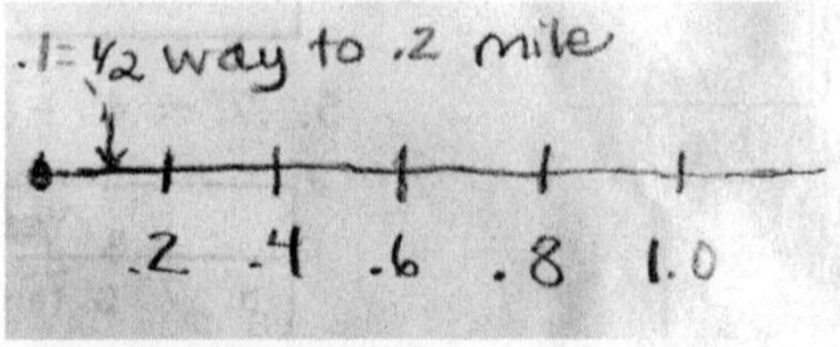

2. a. Yes; possible sketch:

2. b. Yes; possible sketch:

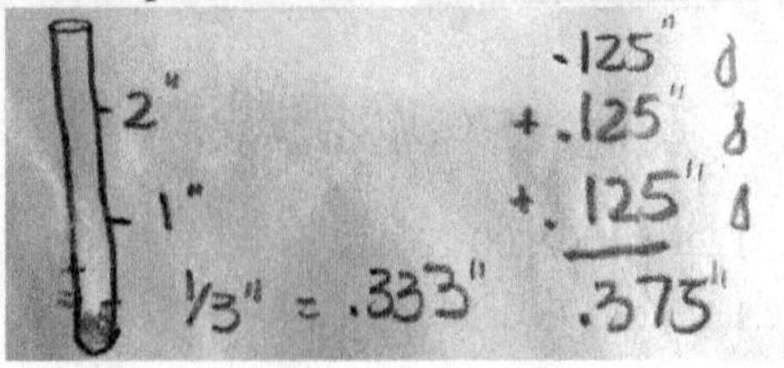

3. a. No; possible sketch:

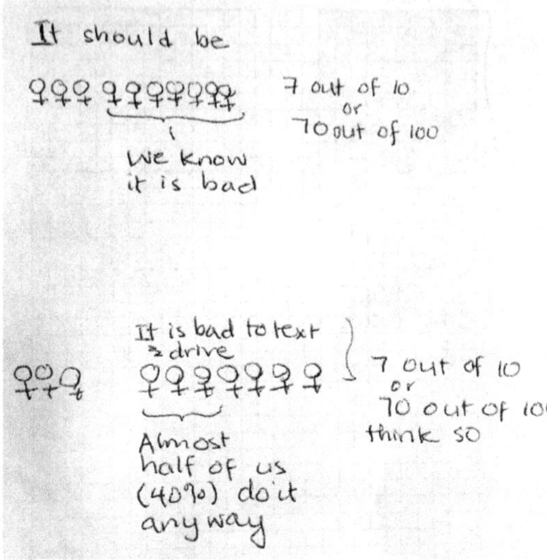

3. b. Yes; possible sketch:

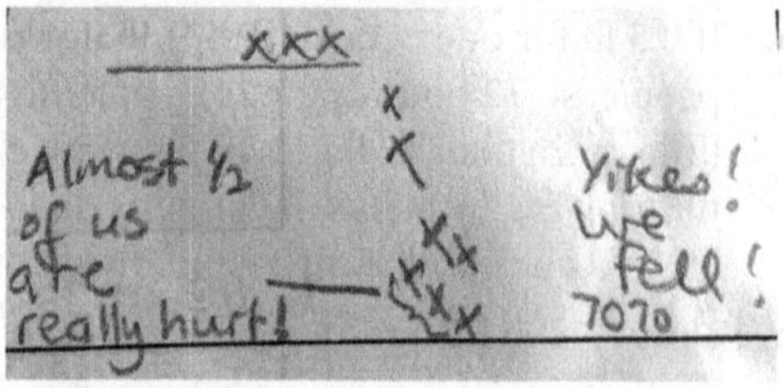

4. Yes; possible sketch:

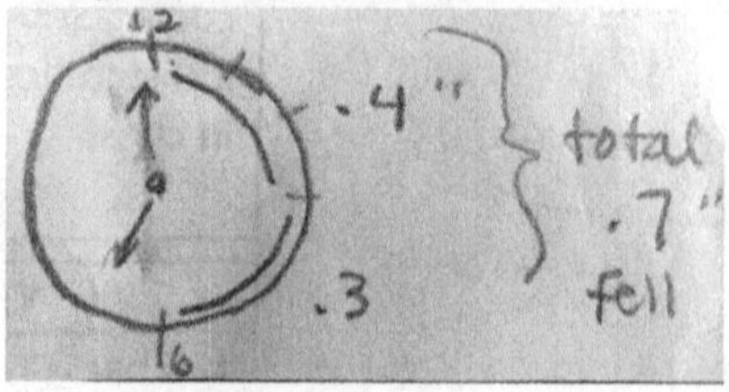

5. Yes; possible sketch:

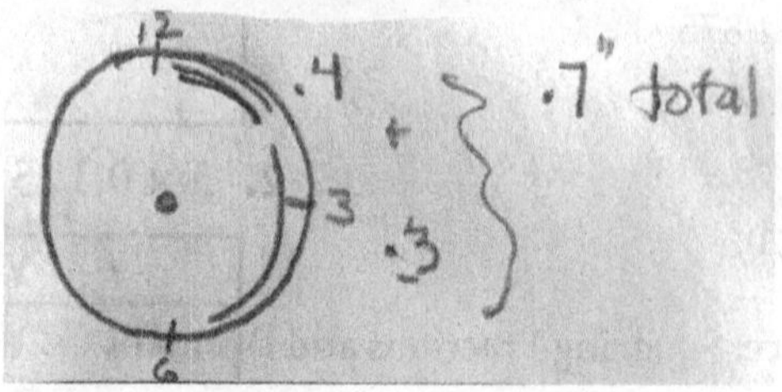

Practice: Estimate Answers to Decimal Multiplication

Part 1

1. 0
2. 0.5
3. 0.05
4. 1.5
5. 0.1

Part 2

1. a. under $33, or $29.

b. Round 2.75 yards to 3 and $10.75 to $11.00, or take $10.75 about two and three-quarter times, so 10.75 + 10.75 + 7.5. To confirm, try it another way.

2. a. around 42 seconds

b. 400 is 4 times as far so 4 × 10 would be 40. Since 0.6 is a little more than 1/2, 4 halves would be 2 which would add another 2 seconds for 42 seconds. To confirm, round 42 to 40 and divide by 4—yes, back to around 10 seconds.

3. a. about 12 pounds
   b. If 0.5 lb for one person, then 1 lb. for two people, so 12 pounds for 24 people, so a little more than 12 lbs or 12.5; or since multiplying by half is the same as dividing by 2, I can round 25 to 24 and divide by 2.
4. a. about 3,200 meters
   b. multiplied 400 × 8

Part 3

1. 0.125
2. 51.5
3. 0085.0
4. 0.0850
5. .00850
6. 001.0
7. 12.875
8. 8.651
9. 198.973
10. .708

Practice: Squaring Fractions and Decimals

1. 1/4
2. 0.0025
3. 87 1/9
4. 6.25
5. 0.04
6. 0.0004
7. 33.64
8. 25/36
9. less than 1
10. greater than 1.
11. Answers will vary. Possible answer: The amount of decimal places is doubled when squared. For example, tenths (one decimal place) becomes hundredths (two decimal places) when squared; and hundredths (2 decimal places) becomes ten thousandths (4 decimal places) when squared.

Practice: Using an Area Model for Multiplication

1. a. Possible answer: It looks like we are splitting up the second rectangle into two parts, but the total areas of the two rectangles will equal the area of the first rectangle.
   b. 7 × 6 = 5 × 6 + 2 × 6

2. a.

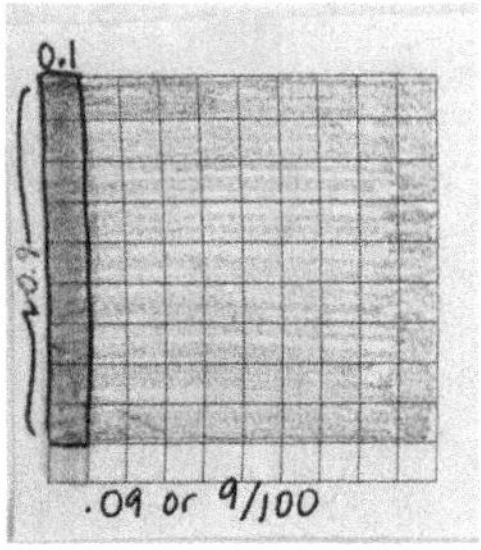

b.

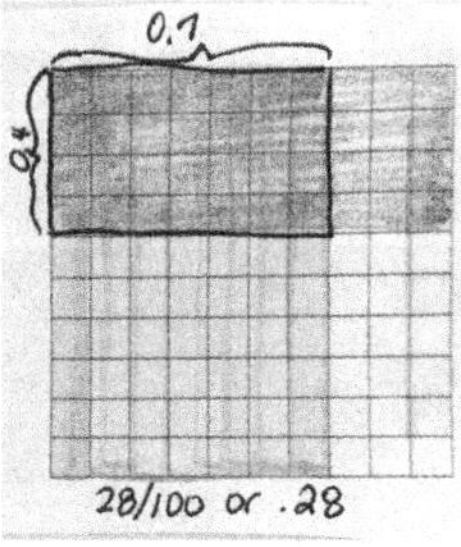

c.

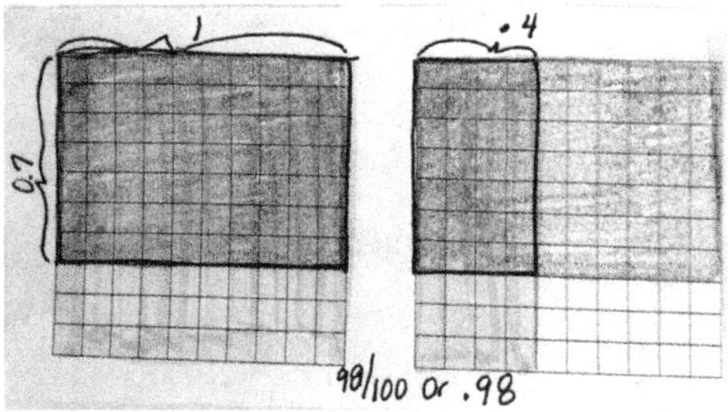

d. 1.2 or 120/100

e. 1.75 or 175/100

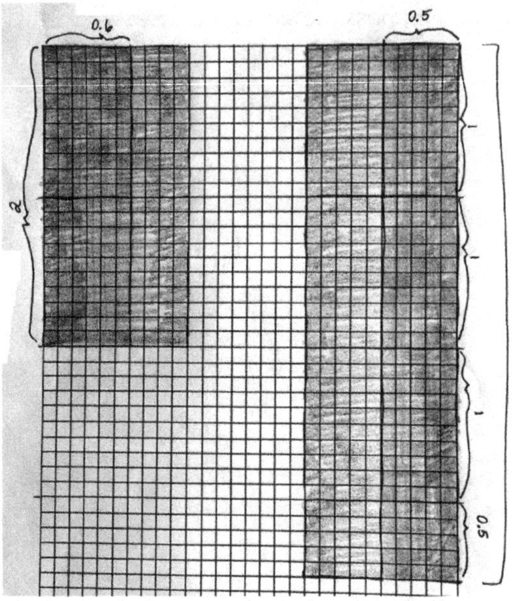

Calculator Practice: Decimals, Decimals

Answers will vary depending on choices, but estimation plays into picking factors that will be choices on the table.

Test Practice

1. (c)
2. (a)
3. (b)
4. (b)
5. (a)
6. 93 1/2 square inches

Lesson 10: Dividing Decimals

Activity 1: What Is the Message?

1. a-b. Depends on what the student selects.

c. Answers will vary. Possible answer: There isn't a whole 5 in 3.25; just part of it.

2. a-b. Depends on what the student selects.

c. Answers will vary. Possible explanation: "There are six and a half halves in 3.25," and this could be shown on the number line, in circles or rectangles, or on the 100-grids; or even using money.

3. a-b. Depends on what the student selects.

c. Answers will vary. Possible representation: the 100-grid with 12.5 squares shaded in to show that there is nowhere near one 100 in 12.5.

4. a-b. Depends on what the student selects.

c. Answers will vary. Possible representation: the 100-grid and shade groups of 12.5 squares to show that there are 8 of them; or they may use money and dole out $12.50 to each person, to find out how many people will receive $12.50, and realize 8 people each received $12.50.

5. In Problems 1 and 2, the total to be divided is the same, but the amount to be divided by is different, 5 groups in Problem 1 and in Problem 2, 0.5 group. Also Problem 2 yields a bigger amount than Problem 1.

6. In Problems 3 and 4, the numbers used in each problem are the same but they are reversed; the solutions are reciprocals of each other: 1/8 and 8.

Activity 2: Methods for Dividing Decimals—They Have to Make Sense!

Part A: Whole Number Divisors

1. a. Answers will vary; a good estimate is around 3 pounds

b. Methods may vary, especially for students from other countries.

c. 2.88

d. Answers will vary; U.S. students may write the following: first place the decimal point above the division sign ($\overline{)\quad}$) in the same place as the decimal you are dividing. Then divide 3 into 8. It goes 2 times, so place the 2 above in front of the decimal. Multiply 3 by 2 and place the 6 under the 8 and subtract to get 2. Next bring the 6 down next to the 2 that was a result of the subtraction. Divide 3 into 26. It goes 8 times. Bring the 8 above the division side to the right of the decimal point. Multiply 8 by 3 for 24. Subtract 24 from 26 for 2. Bring down the 4 from above. Divide 3 into 24. It goes 8 times so place the 8 above the division symbol next to the 8. Multiply 8 by 3 for 24. Subtract 24 from 24 for 0. The answer is 2.88.

e. Answers will vary, especially for a class with ELL students or students who were taught division a different way from the U.S. standard algorithm.

f. The procedures will work, but watch for students who forget about zero placeholders. For 0.156 ÷ 3 many students will incorrectly think the answer is 0.52 when the answer is 0.052. The same is true for 1.5 ÷ 30, students may think the answer is .5 when the answer is 0.05.

Part B: Dividing by a Decimal

1. a. Answers will vary, a good estimate is 16

b. Procedures may vary; especially for students from other countries

c. 16.6

d. Answers will vary; for the U.S. standard algorithm, the process would be to multiply the dividend by 10. Divide 5 into 83. 5 divides into 8 one time. Place the 1 above the division sign above the 8, multiply 1 by 5 to get 5. Subtract 5 from 8 for 3. Bring down the 3 for 33. Divide 33 by 5 for 6.

Place the 6 above the division sign next to the 1. Multiply 6 by 5 for 30 and subtract 30 from 33 for 3. Place a decimal point next to the 6 above the division sign. Add a zero to the 3 for 30. Divide 30 by 5 for 6. Place the 6 to the right of the decimal point above the division symbol. Multiply 6 by 5 for 30. Subtract 30 from 30 for zero. The answer is 16.6.

e. Answers will vary; especially for a class with ELL students or students who were taught division a different way from the U.S. standard algorithm.

f. Watch for misplaced decimal points. The answer to the first example is 5.2, the answer to the second is 500. Students may be surprised that the answer is larger.

Activity 3: Weekly Expenses

1. Some months have 5 weeks while other months have 4 weeks. The average is around 4.3.
2. (a), (b), and (c)

Category	Average Monthly Expenses	Estimated weekly expenses	Weekly expenses to report to bank
Rent	$1,000	$250	$232.56
Cell phone	$69.15	$17	$16.08
Heat (oil)	$113.06	$25	$26.29
Electricity	$40.32	$10	$9.38

d. Answers will vary. For example, some may say that they rounded to a value that was easy to divide by 4.

Practice: Four Ways to Write Division

2.	8)0.6	0.6 ÷8	0.6/8	0.6 divided by 8
3.	7.2)1.24	1.24÷7.2	1.24/7.2	1.24 divided by 7.2
4.	4)0.8	0.8÷4	0.8/4	0.8 divided by 4
5.	3)1.5	1 1/2 ÷3	1 1/2 /3	1 1/2 divided by 3
6.	10)0.25	1/4 ÷10	1/4 / 10	1/4 divided by 10
7.	10)20.3	20.3 ÷10	20.3/10	20.3 divided by 10
8.	10.5)1	1÷10.5	1/10.5	1 divided by 10.5

Practice: Target Practice, 0.1, 0.01, 100

1. Round 1 answers will vary; the closest value is 80.

   Round 2 answers will vary; the closest value is 420.

   Round 3 answers will vary; the closest value is 2,496.

2. a. 800

   b. 4,200

   c. 24,960

   A pattern is to divide by a number with the same digits in place values 100 times larger (or two spaces to the left) than the divisor.

3. a. 2.5

   b. 0.015

   c. 0.001
   A pattern is to divide by a number with the same digits in place values 100 times smaller (or two spaces to the right) than the divisor.

Practice: Which is Not the Same

1. c
2. a
3. c
4. b
5. c
6. d

Practice: Where's the Point?

1. 10.389724
2. 225.8
3. 103.89724
4. 1.0389724
5. 00.650
6. 0.0650
7. 1.00
8. 5.625
9. .5625
10. 15.75

Practice: Multiplication and Division Patterns

1.

Original amount	Multiply by 2	Divide by 2	Multiply by 1/2	Divide by 1/2
b. 26	26 × 2 = 52	26 ÷ 2 = 13	26 × 1/2 = 13	26 ÷ 1/2 = 52
c. 0.5	0.5 × 2 = 1.0	0.5 ÷ 2 = 0.25	0.5 × 1/2 = 0.25	0.5 ÷ 1/2 = 1.0
d. 12 1/2	12 1/2 × 2 = 25	12 1/2 ÷ 2 = 6 1/4	12 1/2 × 1/2 = 6 1/4	12 1/2 ÷ 1/2 = 25
e. 8.9	8.9 × 2 = 17.8	8.9 ÷ 2 = 4.45	8.9 × 1/2 = 4.45	8.9 ÷ 1/2 = 17.8

2.

Original amount	Multiply by 4	Divide by 4	Multiply by 1/4	Divide by 1/4
b. 26	26 × 4 = 104	26 ÷ 4 = 6.5	26 × 1/4 = 6.5	26 ÷ 1/4 = 104
c. 0.5	0.5 × 4 = 2.0	0.5 ÷ 4 = 0.125	0.5 × 1/4 = 0.125	0.5 ÷ 1/4 = 2.0
d. 12 ½	12 1/2 × 4 = 50	12 1/2 ÷ 4 = 3 1/8	12 1/2 × 1/4 = 3 1/8	12 1/2 ÷ 1/4 = 50
e. 8.9	8.9 × 4 = 35.6	8.9 ÷ 4 = 2.225	8.9 × 1/4 = 2.225	8.9 ÷ 1/4 = 35.6

Practice: Division Patterns

1. 8 ÷ 5 = 1.6; 7 ÷ 5 = 1.4; 6 ÷ 5 = 1.2;
   5 ÷ 5 = 1, 4 ÷ 5 = 0.8; 3 ÷ 5 = 0.6; 2 ÷ 5 = 0.4; 1 ÷ 5 = 0.2
2. Answers will vary. The quotient decreases by 0.2 each time because 1/5 = 0.2. In terms of connecting to dividing by 10, dividing by 5 is double what dividing by 10 is because 1/10 doubled is 1/5.

Practice: Think Metric

1. a. 10 ÷ 2.5 = 4

   b.

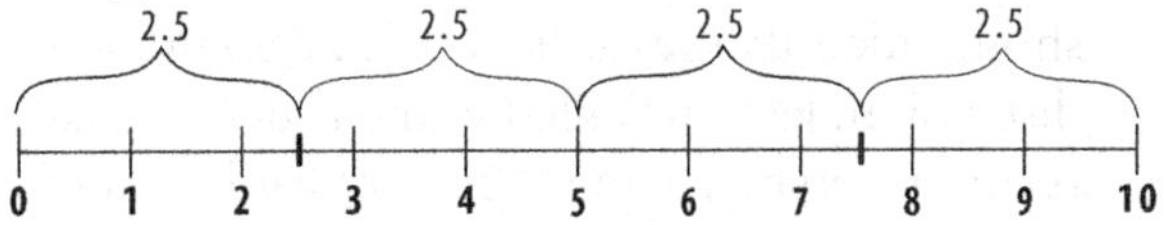

2. a. 3 ÷ 0.1 = 30

   b.

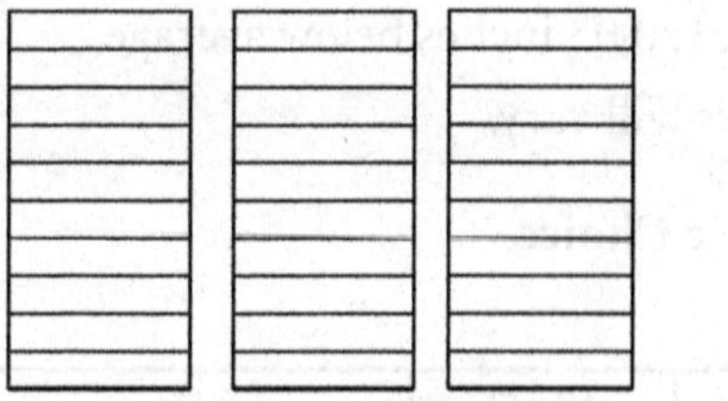

3. a. 16 ÷ 3.5 = 4.57, or 4 portions with some left over

   b.

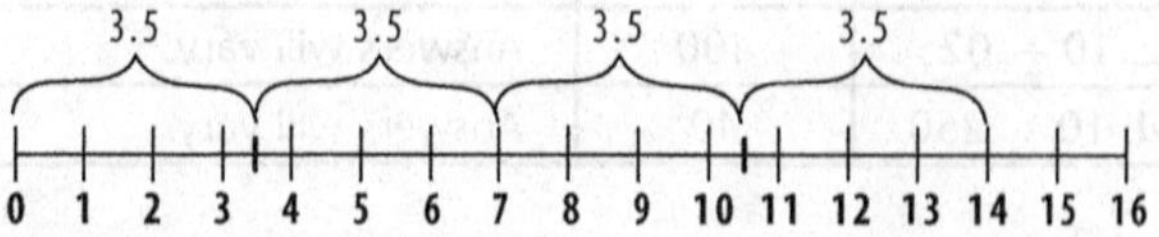

4. a. 5 ÷ 0.75 = 6.67, or 6 with 2/3 left over or 0.667

   b.

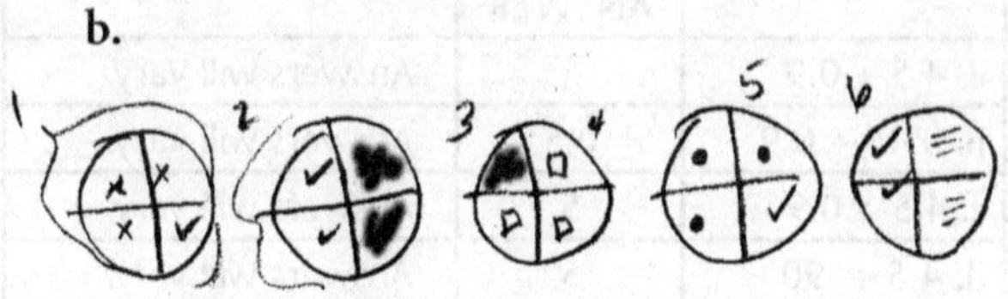

Calculator Practice: Decimal Division

A Thoughtful Estimate	The Exact Answer
1. 37	34
2. 0.25	0.2379
3. 20	19.3
4. 100	86.5217391
5. 34	33.6
6. 8.4	8.4
7. 0.6	0.66
8. 1	0.92307692
9. 3	2.24

Calculator Practice: Way Under Average?

1. Average is 193 ÷ 4 = 48.25; and – 15 = 33.25 minutes
2. 11.9 million ÷ 6 = 1,983,333; and – 1.1 million = $883,333
3. 21.64 ÷ 6 = 3.6066666; and – 3.59 = 0.016; she paid almost 2¢ less than the average
4. Students might compare each month or compare the winter averages.

Dec. 2020: 0.3 inches below average
Jan. 2021: 0.26 inches below average
Feb 2021: 0.03 inches below average

5. Answers will vary.

Practice: Free Choice

Set 1

	WRITE YOUR ANSWER	HOW DID YOU SOLVE IT?
a. 10 ÷ 2.5	4	Answers will vary.
b. 10 ÷.25	40	Answers will vary.
c. 10 ÷ .025	400	Answers will vary.
d. 10 ÷ .250	40	Answers will vary.

Set 2

	WRITE YOUR ANSWER	HOW DID YOU SOLVE IT?
a. 4.5 ÷ 0.9	5	Answers will vary.
b. 0.45 ÷ 0.9	0.5	Answers will vary.
c. 4.5 ÷ 0.9	5	Answers will vary.
d. 4.5 ÷ .90	5	Answers will vary.
e. 0.045 ÷ 0.9	0.05	Answers will vary.

Set 3

	WRITE YOUR ANSWER	HOW DID YOU SOLVE IT?
a. 0.42 ÷ 6	0.07	Answers will vary.
b. 0.042 ÷ 6	0.007	Answers will vary.
c. 4.2 ÷ 6	.7	Answers will vary.
d. 42 ÷ 6	7	Answers will vary.

4. Easiest to do in your head: answers will vary.

Extension: Geometric Formulas

1. a. more than 1 1/2 miles
   b. Answers will vary.
2. .795 of a mile
3. 3.18 miles
4. **a.** 1.75 miles
   b. 3.5 – 3.18 = .32 of a mile, a big deal; swimming takes a lot of energy.

Test Practice

1. (d)
2. (b)
3. (b)
4. (d)
5. (b)
6. 45.6

Lesson 11: Applying Decimal Learning

Activity 1: Number of the Day

Answers will vary.

Activity 2, Project 1: Applying for Life Insurance

1. Test results **(c)** and **(d)** would be flagged.
2. Student answers will vary. Here is an example:

Blood Test	
Test Name	*Test Results of Life Insurance Applicant*
Cholesterol (in mg/dl)	136
LDL/HDL Ratio (in mg/dl)	7.01
Glucose (in mg/dl)	130
Fructosamine (in mmol/l)	2.0
Albumin (in g/dl)	3.5
Creatinine (in mg/dl)	1.8
Total Protein (in g/dl)	6.5
Urinalysis	
Adult. Creatinine (in mg/dl)	265.1
Protein/Creatinine R	0.08
Adulterant PH	4.9

3. Some students may decide to offer the policy although she is outside the acceptable range for three tests, while others may decline her application or decide to charge her more than $23 per month. Any of these options are acceptable for this task, as long as they refer to the flagged tests and demonstrate their ability to order decimal amounts. Students should identify the outliers (red flags) in their discussion. Jasmine failed in the Fructosamine reading (too high), the Albumin reading (too low), and the Protein/Creatinine reading (too high).

Activity 2, Project 2: The Cereal Box Project

1. Answers will vary depending on the box chosen. Below is an example of a Raisin Bran box:

Dimension	Measurement (cm)
a. Height	26.4
b. Width	16.5
c. Depth	4.8

2. Answers will vary depending on the box chosen. Below is an example for the cereal and wrapper:

Dimension	Measurement (cm)
a. Height	22.9
b. Width	15.2
c. Depth	4.1

3. Answers will vary but should be larger than the cereal and wrapper and smaller than the original box. See possible example.

Dimension	Measurement (cm)
a. Height	24
b. Width	15.5
c. Depth	4.5

4. Answers will vary according to the values chosen. Make sure students remember how to find surface area. For the example above, it would be 1,283.0 cm^2 for the cereal box minus 1,099.5 cm^2 for the new box for a savings of 183.5 cm^2.

Activity 2, Project 3: No Wonder It's Noisy!

1. Disagree. Higher values absorb more sound. The painted brick absorbs at most 0.02 of the sound which is lower than glass at 0.05 to 0.10.
2. The fiberglass, 3 1/2 inch batt.
3. Some suggestions might be to use heavy carpet on foam rubber or concrete or cork floor tiles. Another suggestion might be cork wall tiles.
4. Carlos could use cork wall tiles or heavy weight drapery and put in heavy carpeting with rubber matting to help with the noise.
5. Answers will vary.

Activity 2, Project 4: The Stock Market

1. $11,880; $12,355.20
2. $70
3. a. $1.01; 5%
   b. $100
   c. $125
4. a. Disagree. 1% is 0.01 and a half of that is 0.005, or 0.5% so it would be up $50.93 × 0.5%.
   b. $51.18
   c. $250
5. Answers will vary depending on what stock the student chooses.

Closing the Unit: Put It Together

Activity 1: Self Assessment

Answers will vary.

Activity 2: Final Assessment

Task 1: Number of the Day

Answers will vary.

Task 2: Vegetable Sale

	Income
1. 5%	$12.50
2. 10%	$25.00
3. 20%	$50.00
4. 25%	$62.50
5. 40%	$100.00

Task 3: Frames

1. 6.25 × 4 = (6 × 4) + (.25 × 4) = 24 + 1 = 25"
2. 5 3/8 × 6 = (5 × 6) + (3/8 × 6) = 30 + 18/8 = 32 1/4"

Task 4: Write It in Numbers

1. .03
2. 750,000
3. 250,000
4. 75,000
5. 9.52
6. 4.021
7. 30.21
8. 2,021,000

Task 5: More Than, Less Than, or the Same As

1. less than
2. more than
3. the same as
4. less than

Task 6: Mole Removal

On the number line, the numbers should be ordered (from left to right): 0.08. 0.09, 0.16, 0.91, 4.5, 9. The 0.08, 0.09, and 0.16 will all be clustered at the far left end of the line; 0.91 should be close to 1 mm; 4.5 should be close to the middle, and 9 mm toward the right end.

Task 7: Surgical Cuts

1. 0 .08 mm, 9 mm
2. 9 mm, 0.91 mm
3. 0.08 mm, 0.16 mm or 4.5 mm, 9 mm

Task 8: Place the Decimal Point

1. 1.16
2. 9.58
3. 1.92
4. 0.012

Task 9: S&P Increase

1. 7.5 points
2. Answers will vary, but one possible way is:

   $\frac{1}{2}$ of 1% of 1500

   10% of 1500 = 150

   1% of 1500 = 15

   $\frac{1}{2}$ of 1% of 1500 = $\frac{1}{2}$ of 15 = 7.5

Task 10: A Smaller Box

1. 6.3 cm in length
2. 4.05 cm in width
3. 1.575 cm in height

Task 11: Aluminum Cans

About 66 to 67 layers.

Task 12: Comparing Thickness

Answers may vary; students might specify that the 2013 can is 0.0103 inches thinner, or they might say the new cans are about 3 times thinner than they were in the 1970's.

Sources and Resources

Mathematics Education

The National Council of Teachers of Mathematics (NCTM) publishes several excellent resources.

- *Principles to Actions - Ensuring Mathematical Success for All, 2014.*
- *Principles and Standards for School Mathematics,* 2000.
- The NCTM journals: *Mathematics Teaching in the Middle School, Mathematics Teacher, Mathematics Teacher Educator,* and *Journal for Research in Mathematics Education.*
- *Developing Number Sense in the Middle Grades, The Addenda Series, Grades 5–8,* 1991.
- *Historical Topics for the Mathematics Classroom,* 1989.

For more information on NCTM resources, visit http://www.nctm.org

Boaler, Jo. *What's Math Got to Do With It?* New York, NY: Penguin Books, 2015.

Boaler, Jo. "Memorizers Are the Lowest Achievers and Other Common Core Math Surprises." *The Hechinger Newsletter.* The Hechinger Report. 7 May 2015. <https://hechingerreport.org/memorizers-are-the-lowest-achievers-and-other-common-core-math-surprises/>

Burns, M. "Introducing Multiplication of Fractions: A Lesson for Fifth and Sixth Graders." *Math Solutions Newsletter.* Math Solutions. Issue 12, Winter 2003-2004. <http://www.mathsolutions.com/documents/0-941355-64-0_L.pdf>

Cengiz, N., and M. Rathouz. "Take a Bite out of Fraction Division." *Mathematics Teaching in the Middle School* 17.3 (2011): 146-53.

Clarke, D., A. Roche, and A. Mitchell. "10 Practical Tips for Making Fractions Come Alive and Make Sense." *Mathematics Teaching in the Middle School.* 13. 7 (2008): 372-80.

Driscoll, M. *Fostering Algebraic Thinking: A Guide for Teachers, Grades 6-10.* Portsmouth, NH: Heinemann, 1999.

Driscoll, M. *The Fostering Algebraic Thinking Toolkit* Portsmouth, NH: Heinemann, 2001.

Fendel, D., D. Resek, L. Alper, & S. Fraser. *Interactive Mathematics Program.* Berkeley, CA: Key Curriculum Press, 1997.

Gregg, J., and D. Gregg. "Measurement and Fair-Sharing Models for Dividing Fractions." *Mathematics Teaching in the Middle School* 12.9 (2007): 490-96.

Huinker, D. "Examining Dimensions of Fraction Operation Sense." *Making Sense of Fractions, Ratios, and Proportions, 64th Yearbook.* Eds. George Bright and Bonnie Litwiller. Reston, VA: NCTM, (2002) 72-78.

Lappan, G., J. Fey, W. Fitzgerald, S. Friel, & E. Phillips. *Connected Mathematics Series.* Parsippany, NJ: Dale Seymour Publications, Division of Pearson Education, 1998.

Markovitz, Z. and J.T. Sowder. "Students' understanding of the relationship between fractions and decimals." *Focus on Learning Problems in Mathematics* 13.1 (1991): 3–11.

Morrison, P., and P. Morrison. *Powers of Ten.* New York: Scientific American Library, 1994.

Moschkovich, J. *Mathematics, the Common Core, and Language: Recommendations for Mathematics Instruction for ELs Aligned with the Common Core.* 2012. <https://files.eric.ed.gov/fulltext/ED584919.pdf>.

National Research Council. *Adding It Up: Helping Children Learn Mathematics.* J. Kilpatrick, J. Swafford, and B. Findell (Eds.). Mathematics Learning Study Committee, Center for Education, Division of Behavioral and Social Sciences and Education. Washington, DC: National Academy Press, 2001.

Rasmussen, S. *Key to Fractions.* Columbus, OH: McGraw-Hill Education, 1993.

Reys, B.J. "Teaching Computational Estimation: Concepts and Strategies." *Estimation & Mental Computation-1986 Yearbook.* Eds. Harold L. Schoen and Marilyn J. Zweng. Reston, VA: NCTM, 1986. 31-44.

Reys, B.J., O. Kim, and J. M. Bay. "Establishing Fraction Benchmarks." *Mathematics Teaching in the Middle School* 4.8 (1999): 530-32.

Russell, S.J., K. Economopoulos, L. Wittenberg, et al. *Investigations in Number, Data, and Space ®, Second Edition.* Glenview, IL: Pearson, 2012.

Russell, S.J., D. Schifter, and V. Bastable. *Connecting Arithmetic to Algebra: Strategies for Building Algebraic Thinking in the Elementary Grades.* Portsmouth, NH: Heinemann, 2001.

Schifter, D. "Examining the Behavior of Operations: On Noticing Early Algebraic Ideas." *Mathematics Teacher Noticing: Seeing through Teachers' Eyes.* Eds. Myriam Sherin, Vicki Jacobs and Randy Philipp. New York: Routledge, 2010. 204-20.

Seeley, C.L. *Smarter Than We Think - More Messages About Math, Teaching, and Learning in the 21st Century.* Sausalito, CA: Math Solutions, 2014.

Siegler, R., T. Carpenter, F. Fennell, D. Geary, J. Lewis, Y. Okamoto, L. Thompson, and J. Wray. *Developing Effective Fractions Instruction for Kindergarten through 8th Grade: A Practice Guide* (NCEE #2010-4039). Washington, DC: National Center for Education Evaluation and Regional Assistance, Institute of Education Sciences, U.S. Department of Education, 2010. <https://ies.ed.gov/ncee/wwc/Docs/PracticeGuide/fractions_pg_093010.pdf>.

Storeygard, J. *My Kids Can: Making Math Accessible to All Learners, K-5.* Portsmouth, OR: Heinemann, 2007.

Van de Walle, J., K. Karp, J. Bay-Williams . *Elementary and Middle School Mathematics: Teaching Developmentally.* 8th Edition. New York: Pearson, 2012.

Mathematics and Numeracy Education for Adults

Curry, D., M.J. Schmitt, and S. Waldron. *A Framework for Adult Numeracy Standards: The Mathematical Skills and Abilities Adults Need to Be Equipped for the Future.* Boston: World Education, 1996. (This framework was developed by members of the Adult Numeracy Network.).

Givvin, K. B., J. W. Stigler, and B. J. Thompson. "What Community College Developmental Mathematics Students Understand About Mathematics, Part 2: The Interviews." *MathAMATYC Educator 2.3* (2011): 4-18.

Massachusetts Department of Education, Adult and Community Learning Services. *Massachusetts Adult Basic Education Curriculum Frameworks for Mathematics and Numeracy.* Malden, MA: Massachusetts Department of Education, 2005.

National Center on Education and the Economy. *What Does It Really Mean to Be College and Work Ready?* Washington, DC: 2013. <http://www.ncee.org/wp-content/uploads/2013/05/NCEE_ExecutiveSummary_May2013.pdf>.

Stein, S. *Equipped for the Future Content Standards: What Adults Need to Know and Be Able to Do in the Twenty-First Century.* ED Pubs document EX0099P. Washington, DC: National Institute for Literacy, 2000.

Stigler, J. W., K. B. Givvin, and B. J. Thompson. "What Community College Developmental Mathematics Students Understand About Mathematics." *MathAMATYC Educator 1.3* (2010): 4-16.

Web Sites

http://www.funbrain.com/fract/

https://www.mathlearningcenter.org/apps/pattern-shapes

https://www.mathlearningcenter.org/apps/fractions

https://toytheater.com/fraction-bars/

Made in the USA
Monee, IL
08 November 2026

68032039R00122